W9-AAH-759

Robust Engineering

Robust Engineering

Genichi Taguchi
Subir Chowdhury
Shin Taguchi

Feb.2007
@ DCX

McGraw-Hill
New York · San Francisco · Washington, D.C
Auckland · Bogotá · Caracas · Lisbon · Madrid
Mexico City · Milan · Montreal · New Delhi
San Juan · Singapore · Sydney · Tokyo · Toronto

Library of Congress Cataloging-in-Publication Data

Taguchi, Genichi, 1924 –
Robust engineering / Genichi Taguchi, Subir Chowdhury, Shin Taguchi
p. cm.
ISBN 0–07–134782-8
1. Quality control Case studies. 2. Production engineering Case studies. 3. Design, Industrial Case studies. 4. Taguchi methods (Quality control) Case studies. I. Chowdhury, Subir. II. Taguchi, Shin. III. Title.
TS156.T335 1999
658.5—dc21 99–41786
CIP

Copyright © 2000 by American Supplier Institute, Inc., Genichi Taguchi, Subir Chowdhury, Shin Taguchi. All rights reserved. Printed in the United States of America. Except as permitted under the United States Copyright Act of 1976, no part of this publication may be reproduced, stored in a retrieval system, or transmitted, in any form or by any means, electronic, mechanical, photocopying, recording, or otherwise, without the prior written permission of the publisher. Visit the authors at www.robustengineering.net

4 5 6 7 0 DOC/DOC 0 4

ISBN 0–07–134782–8

The sponsoring editor for this book was Linda Ludewig and the production supervisor was Pamela A. Pelton.

It was set in Garamond by IDS Design.

Printed and bound by R.R. Donnelley & Sons, Inc.

McGraw-Hill books are available at special quantity discounts to use as premiums and sales promotions, or for use in corporate training programs. For more information, please write to the Director of Special Sales, McGraw-Hill, Professional Publishing, Two Penn Plaza, New York, NY 10121-2298. Or contact your local bookstore.

This book is printed on recycled, acid-free paper containing a minimum of 50% recycled, de-inked fiber.

Information contained in this work has been obtained by the McGraw-Hill Companies, Inc. ("McGraw-Hill") from sources believed to be reliable. However, neither McGraw-Hill nor its authors guarantee the accuracy or completeness of any information published herein and neither McGraw-Hill nor its authors shall be responsible for any errors, omissions, or damages arising out of use of this information. This work is published with the understanding that McGraw-Hill and its authors are supplying information but are not attempting to render engineering or other professional services. If such services are required, the assistance of an appropriate professional should be sought.

To our colleagues, friends, and families
for their continuous support

ACKNOWLEDGMENTS

The authors gratefully acknowledge the efforts of all who assisted in the completion of this book:

To all of the contributors and their organizations for sharing their successful case studies.

To the American Supplier Institute and its employees, especially Jim Wilkins and Alan Wu, for promoting Robust Engineering.

To Yuin Wu, a long time friend of Dr. G. Taguchi who has assisted translating most of the international case studies of this book, for his dedication on Robust Engineering.

To the Japanese Society of Quality Engineering for letting us use some case studies originally published by JSQE.

To Rajesh Jugulum, a true researcher in Robust Engineering for his constant helping hands on this project.

To Glenn Davis of ITT Industries for his valuable suggestions on the book's manuscript.

To Mike Hayes, our publisher at McGraw-Hill for his vision and encouragement from day one.

To Linda Ludewig, our editor at McGraw-Hill for her continuous support.

To Melissa Verduyn for her dedicated effort on design and graphics.

CONTENTS 

To compete successfully in the global marketplace, organizations must have the ability to produce a variety of high-quality, low-cost products that fully satisfy customers' needs. Tomorrow's leaders will be those companies that make fundamental changes in the way organizations develop technologies, and design and produce products. American Supplier Institute, Inc. has been improving the competitive position of organizations around the globe by introducing new and powerful tools like Robust Engineering™.

Robust Engineering is the creation of world-renowned engineering genius and quality guru – Dr. Genichi Taguchi. In this pioneering work, authors define this new method and demonstrate how these techniques can be applied, and how to build into manufacturing organizations the flexibility that minimizes product development cost, reduces product time-to-market, and increases overall productivity. Thousands of companies throughout the world have been using Dr. Taguchi's methodologies for the past 30 years with excellent results. Some organizations have integrated this new, powerful methodology in their corporate culture. The benefits these organizations have achieved are monumental. There are a multitude of companies today, such as – Ford Motor Company, Xerox Corporation, ITT Industries, Minolta, Goldstar, JPL, Nissan, Fuji-Xerox, Case Corporation – that are implementing Robust Engineering in their everyday activities while developing innovative and outstanding products for the current and future marketplace. In this groundbreaking book, authors have introduced 18 of the best successfully implemented case studies from organizations around the world.

PURPOSE OF THIS BOOK

Prior to this, there were no books available in any language which presented a worldwide collection of the BEST PRACTICES of Robust Engineering. In the organizations where Robust Engineering is extremely successful, management has an understanding of the significant impact of the applications to productivity and cost. In these firms the support for implementation is highly encouraged or mandatory. This quality-building philosophy is elegantly simple but at the same time an extremely powerful tool.

In Japan, Robust Engineering is extremely popular. Japanese companies are committed to the utilization of these tools because of management's understanding that these simple yet powerful tools enhance product quality and workforce productivity resulting in a higher level of return on investment ultimately increasing "the bottom line". In the United States and Europe, hundreds of organizations have been using this methodology and saving millions of dollars. Yet still today, thousands of American, European, Asian, and Australian organizations are not utilizing these powerful methods because of the lack of management understanding and misconceptions about its complexity.

Importance of Actual Case Studies

This book outlines a series of successful case studies profiling organizations who are on the cutting edge of technology and continue to lead the way with the aid of Robust Engineering. This book also provides a wide range of discipline specific applications so organizations can more easily understand how to implement these methods in their technologies.

About the Book

It is the authors' intention to create an atmosphere of excitement to management, engineering, academia, and researchers. Chapters One and Two outline a basic explanation of Robust Engineering and an implementation strategy for the method. Chapter Three illustrates a step by step process to understand the Robust Engineering methodology via the case study technique. Chapters Four through Eighteen are case studies from respected companies, contributed directly by members of each organization. Each case study is similar in that they follow the same basic steps of the method, but is unique in the variety of their technique. The final chapter outlines introduces Dr. Taguchi's latest research, the Mahalanobis-Taguchi System (MTS) – a powerful tool for the next millennium.

Intended Audience

This book is for anyone interested in Robust Engineering: engineers (design, product, manufacturing, mechanical, electrical, chemical, electronics, software, process, quality, aerospace etc), all levels of management, research and development personnel, and consultants. This exceptional compilation of case studies makes it ideal as a training and educational guide, as well as serving academic pursuits.

CHAPTER ONE 

Introduction to Robust Engineering Methodology

BACKGROUND AND
INTRO TO ROBUST ENGINEERING

INTRODUCTION TO ROBUST ENGINEERING

This chapter provides an introduction to the methodology of Robust Engineering. As Robust Engineering is based on Taguchi Methods™, it is important to give an overview of Dr. Genichi Taguchi, some insight into the history of the methods development, definitions of the terminology and a description of the basic procedures.

QUALITY ENGINEERING

There are two types of quality:

Type 1: Customer quality (features what customer wants)
Type 2: Engineered quality (problems customer does not want)

Customer quality leads to the size of the market segment. It includes items such as the function itself, features, color and designs. The better the customer quality, the bigger the market size becomes. In order to obtain the market size, the price must be reasonable. Customer quality is addressed during product planning. Robust engineering does not deal with customer quality. Customer quality is extremely important to create a new market.

On the other hand, engineered quality includes defects, failures, noise, vibrations, unwanted phenomena and pollution. Robust engineering is extremely important in winning the bigger market share. While customer quality defines the market size, engineered quality wins the market share within the segment. Robust Engineering deals with engineered quality.

The job of the engineer is to create and develop a system to perform the function required by customer quality. First, several concepts are generated and one or a few concepts go under development. The engineers develop the product so that it performs the intended function under normal conditions. Then the design will be tested under various customer usage conditions. For example, during the 1980s at Bell Laboratories, a new design was tested under 16 customer usage conditions. When the design failed at a certain condition, the design would be changed until it passed all 16 customer conditions. This is not what we call Robust Engineering. This method is called a whack-a-mole engineering.

Robust Engineering optimizes for the robust function. As soon as a concept is selected, numerous design configurations are tried under a few customers' usage conditions to optimize the design for robustness. The design must be optimized for robustness before considering any kind of compensations. Once it is decided that a compensation function be adapted, then the compensation must be optimized for robustness.

It is management's job to change the current product development process to the Robust Engineering process.

Robust Engineering Overview

Robust Engineering is an engineering optimization strategy ideally used for the development of new technologies in the areas of product and process design.

Robust Engineering

- Represents the application of Taguchi Methods at the start of R & D or advanced product/process development activities to optimize performance.
- Concentrates on identifying the "ideal function(s)" for a specific technology or product/process design.
- Concentrates on selectively choosing the best nominal values of design parameters that optimize performance reliability (even in the presence of factors causing variability) at lowest cost.

Robust Engineering was developed by Dr. Genichi Taguchi to provide companies with more efficiency in development leading to a more competitive position. Dr. Taguchi derived this methodology from many years of research and design.

Background

After World War II, the General Headquarters of the Allied forces was established in Tokyo, Japan. At this time, it was discovered that the Japanese telephone system was far below the quality level represented in the United

States and Europe. Reinforcing this observation was the fact that connections routinely took a long time to establish and/or were lost within a short period of time.

Recognizing this, the Allied forces put forth an order to the Japanese government to establish an R & D organization to improve the quality of the communications systems.

In 1950, 26-year-old Dr. Genichi Taguchi joined the Electrical Communication Laboratory (ECL). Having already established himself as a reputable consultant with many successes in the area of industrial experimentation, Dr. Taguchi (along with the section he joined) applied engineering tools to enhance efficiencies in technical activities.

After six years, the ECL successfully completed the development of new phone system components meeting all of the necessary requirements. Thus, as Bell Labs fell behind and struggled to meet requirements, the Japanese telephone companies awarded their business to the team at ECL, resulting in the first true success story by a Japanese company. In fact, Bell Labs never completed the project. It is probably not fair to credit all the success to Dr. Taguchi; however, the ECL had applied certain strategies based on his ideas. Basically, his ideas concentrated on the following questions:

- What information is generated in order to accomplish superior product/process design and meet all requirements at once?
- What should be measured as data in order to generate the best information?
- How should the experimentation be designed?
- How should the data be analyzed?
- How is the validity of a result confirmed?
- How are these methods implemented?

This thinking evolved to become the latest form of Taguchi Methods for Robust Engineering.

ECL was very eager to develop a robust (high quality, reliable and durable) system because they did not manufacture products. ECL designed the product and contracted the manufacturing to other firms from which they purchased the products for lease to the users. ECL, therefore, did not have control over the quality or reliability of the manufacture of the products but ultimately became responsible for their repair/replacement if they failed to work. This greatly affected their bottom line since the expense for repair/replacement came in the form of warranty costs. Thus, the birth of Robust Engineering.

One of the major unique considerations of Dr. Taguchi is that engineers can become more efficient in the evaluation of numerous problems associated with a design by concentrating not on the symptoms of poor function, but on the function itself. Problems are caused due to variability in the design's function. By implementing this philosophy, improving the function minimizes problems.

This type of "thinking" represents one of the most important factors of Robust Engineering. It is also one of the most difficult concepts to appreciate.

What Is Robustness?

Every engineer's dream is to develop a product or process design that exhibits the state of "robustness." However, few can actually make the claim that they know exactly what robustness means.

Webster's dictionary defines "robustness" as being:

- powerfully built, sturdy
- boisterous, rough
- marked by richness and fullness

While these definitions convey what robustness means from a semantics perspective, they do not reflect the concept from Dr. Taguchi's point of view.

Dr. Taguchi defines "robustness" as:

the state where the technology, product, or process performance is minimally sensitive to factors causing variability (either in the manufacturing or user's environment) and aging at the lowest unit manufacturing cost.

Robustness is the goal being sought in robust engineering.

How Is Robustness Measured?

Having defined what robustness is, one may now wonder: How is it measured? When does an engineer know when a design is robust? What measures do typical engineers evaluate to give them indications of robustness? When asked these questions, engineers, more often than not, respond as follows. We measure the robustness of our designs using metrics like:

- Meeting Specifications
- Reliability Data
- Warranty Information
- Cp/Cpk
- Scrap
- Rework
- % Defective
- Failure Rate
- Yield

The problem with relying on these measures to evaluate robustness is that they come too late in the product development cycle, especially if the engineer is working in R & D or advanced engineering environments.

Wouldn't it be great if there existed a more efficient measure of robustness? The good news is . . . there is! It's called the signal-to-noise ratio.

Signal-to-Noise (S/N) Ratio

In fact, Dr. Taguchi states the only good measure of robustness is the signal-to-noise ratio. So what does all of this have to do with the concept of Robust Engineering? EVERYTHING!!

Robust Design

From the description you read on the first page of this chapter, it should become clear how "robustness" and signal-to-noise ratio relate to the concept of robust design. It is the goal of robust design to search for and obtain the state of robustness in product/process designs. Further, it is the signal-to-noise ratio that will be utilized to measure robustness.

Now that the concept of robust design has been defined, and its goals identified, let's consider why the use of this concept is critical to an organization's competitive position.

The Engineered System and Ideal Function

Engineered Systems

One question that is frequently asked is: "Why use robust design?" The answer to this question can be summed in one word . . . efficiency. Robust design enables the engineer to efficiently gather the technological information required to produce high quality, low cost products. Without this structured, engineering-based methodology, the engineer may be forced to utilize other tools/methods that may not even be suited to the role of the engineer.

Engineered systems are man-made systems which use energy transformations to convert input energy into specific, intended output energy by utilizing the laws of physics. These engineered systems are designed to deliver specific results requested by customers.

In every engineered system, there exists some form of perfect or ideal relationship between the input to the system and the output. Robust design seeks to attain that ideal state, referred to as the design's ideal function.

Ideal Function

While every generic system has some sort of input and output, engineered systems transform energy such that the preferred relationship existing between the input energy to the system and the resulting output response is observed.

Achieving efficiency in this transformation of energy – obtaining an ideal relationship between input energy and output response – represents the strategic thinking supporting the ideal function concept. Any engineered system reaches its "ideal function" when all of its applied energy (input) is transformed efficiently into creating desired output energy.

If these were possible, life would be much less stressful for the engineer and the consumer. There would be no energy losses to create symptoms of poor function. As a result, there would be no squeaks, rattles, noise, scrap, rework, quality control personnel, customer service agents, complaint departments, etc.

In reality, nothing functions like this. Every system is less than 100% efficient in its energy transformation. This means that the energy loss goes to create unintended function. Further, the bigger the percentage of energy loss, the bigger the headache.

Signal-to-Noise Ratio

Preceding this, the concept of "robustness" was defined as "representing the state where the product/process design is minimally sensitive to factors causing variability, at the lowest cost." Intuitively, this concept makes sense. The problem surfaces when one thinks of "how it is measured."

Dr. Taguchi states that the only measure of robustness an engineer is required to evaluate is the signal-to-noise ratio. This is particularly true if the engineer's activities are upstream in the R & D and advanced engineering phases of design. Without it, the engineer is forced to evaluate the less efficient, downstream characteristics of quality.

The signal-to-noise ratio is an index of robustness as it measures the quality of energy transformation that occurs within a design. The quality of its energy transformation is expressed as the ratio of the level of performance of the desired function to the variability of the desired function. The higher the ratio, the higher the quality.

Signal-to-noise ratio is used to measure "robustness." The concept of signal-to-noise ratio is shown in Figure 1.

Figure 1: Concept of Signal-to-Noise Ratio

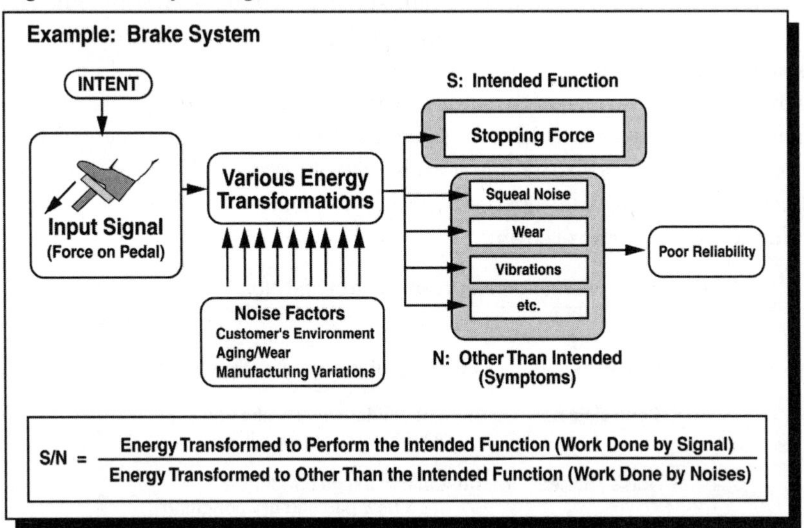

Improving Signal-to-Noise Ratios

In general, there are many strategies and techniques used to increase the value of the S/N ratio. However, one of the most important techniques to increase this value is associated with the engineer's ability to specify the proper nominal values of the design parameter settings (levels) that will make the design robust against noise.

If the engineer does not possess an efficient and effective strategy for testing and identifying the best levels of the design parameters, the search for robustness will be, at the very least, costly, if not impossible. To remedy this, Dr. Taguchi recommends the use of parameter design.

Control factors are any of the design parameters of a system that allow an engineer to freely specify nominal values and maintain cost effectively.

Control factors are used to optimize product/process for robustness.

Control Factors vs. Noise Factors

It is much more productive to do parameter design experimentation at the early stages of a project than later. As a project moves downstream from

R & D through the development cycle to the customer, more and more noise factors enter the picture and the engineer has less freedom to revise the fundamentals of the design. It is more effective to conduct experimentation at the upstream stage when fewer factors have been decided upon, and design changes are less expensive.

Achieving robustness is to take advantage of interactions between control factor and noise factor at the upstream stage in product development. This will prevent unnecessary downstream fire-fighting activities. An example of interaction between control factor and noise factor is shown in Figure 2. It is obvious that the A2 design is more robust than A1; thus A1 has a higher signal-to-noise ratio than A2. Moreover, interaction between control factor and signal factor is also the key to achieve robustness. Chapter 3 and case studies in this book will illustrate the key strategies and thought processes needed to achieve optimization.

Figure 2: Interaction Between Control and Noise

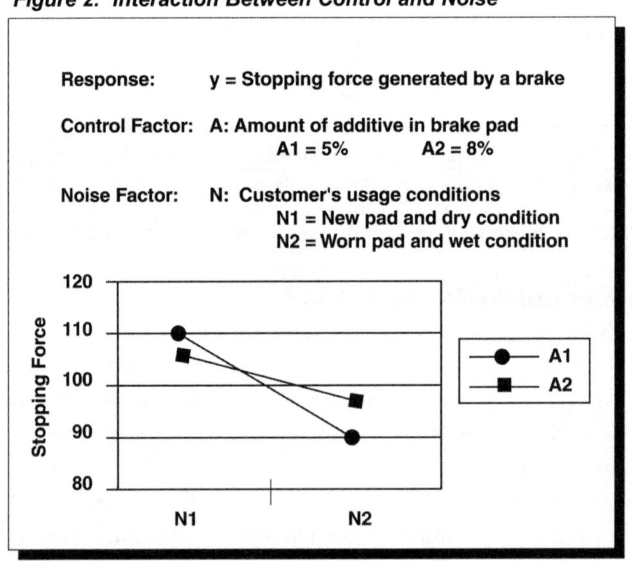

In order to study these interactions, the signal and noise factors are intentionally varied in an experiment. The signal factor must be varied to cover all ranges of operation to assure performance, including future product requirements. A few noise factors are selected and varied in the experiment. This is called a noise strategy such that the robustness can be evaluated for all customers' environmental and aging characteristics. Noise strategies will be illustrated in the case studies.

Two-Step Optimization

Two-step optimization is essential. It is more difficult to reduce variability than to adjust the average response (also called sensitivity) to the target value. When a response represents the energy transformation of the system, it is easy to adjust the sensitivity. In Robust Engineering, two-step optimization is applied.

Step 1: Reduce functional variability
Step 2: Adjust sensitivity

Orthogonal Array

The design space is large. There are tens of control factors within the design space. Engineers must specify the nominal values of control factors, such that the design is optimized for robustness. In order to explore the design space, control factors are assigned an orthogonal array. The weakness of a typical shotgun approach or the one-factor-at-a-time method is that conclusions are drawn from a fixed design configuration. An orthogonal array provides a balanced set of experimentation runs such that the conclusions are drawn in a balanced fashion. When the experiment confirms, there is no severe interaction among control factors to the signal-to-noise ratio; therefore it is very likely that the robustness will be reproduced at the downstream conditions. In Robust Engineering, an orthogonal array L_{12}, L_{18}, L_{36} or L_{54} is used for control factors.

An orthogonal array is used for optimization, i.e., to maximize the signal-to-noise ratio. However, when the objective is simply benchmarking, it is not necessary to use an orthogonal array but instead use the signal-to-noise ratio as the most important evaluation criteria.

Prediction and Confirmation

It is a must to predict the gain in the signal-to-noise ratio and to confirm the gain. It is necessary to complete and rotate the plan-do-check-act cycle.

Strategies of Robust Engineering introduced in this chapter will be illustrated by the case studies in Chapters 3 to 18.

CHAPTER TWO

Robust Engineering Implementation Strategy

EXPERIENCES AND PERSPECTIVES REGARDING THE USE AND INTEGRATION OF ROBUST ENGINEERING

INTRODUCTION

Chapter 2 will introduce the implementation processes of Robust Engineering into the corporate environment and provide the complete and detailed strategies behind the process. This chapter will also show the compatability and the enhancements of any other quality program or process that organization may already have in place. There are those organizations that absolutely thrive on change and embrace the challenges and opportunities that come with change and diversity. Those organizations will have the foresight and the vision to see that the Robust Engineering implementation strategy can be a guide for them. Robust Engineering will ultimately help them to stay on the cutting edge and ahead of their competition by incorporating these processes with what they may already be doing.

ROBUST ENGINEERING IMPLEMENTATION: THE SIX-STEP METHODOLOGY

The six-step implementation process follows as below. It is extremely important that all of these elements be used in the process which makes up the strategies of methodology in the implementation. The number of steps may be added or deleted as needed to complement corporate strengths or to enhance and build in less-developed areas of the organization. Also, without the full support of managers following the implementation strategies for the Robust Engineering process, the organization will not receive the full benefit of this program. Companies who face major challenges and have the greatest needs for a working quality process will definitely have the most to gain from Robust Engineering.

1. Management Commitment
2. Corporate Leader and the Corporate Team
3. Effective Communication
4. Education and Training
5. The Integration Strategy
6. Bottom-Line Performance

MANAGEMENT COMMITMENT

Robust Engineering is one of the most powerful tools for making higher quality products with little or no waste, and making them faster. However, like any new quality process, the buy-in from top and middle management is essential for its implementation. No matter how good or how efficient a program might be, it can be extremely difficult for those individuals at the lower end of the organization if top and middle management are not committed to supporting and making it work. It is essential that management fully understand all of the benefits of the process, and that the program is in no way a threat to their position or to their importance to the organization. Today, everything in any organization is related to the quality of the organization. Everyone wants everything to be more efficient, faster, cheaper and of higher quality. However, the only organizations that will achieve this status will be those that are ready and willing to invest in more advanced and efficient methodology.

Upon completion of all six steps of the implementation process, it is important that everyone in the organization be able to see that the process is fully understood and total commitment is demonstrated by those of higher position. When top management is not only supportive, but encouraging, the rest of the organization will follow suit, and momentum will be gained as it continues to flow throughout the entire organization. The attitude of top and middle management provides the flavor for the process. A negative approach or attitude by the powers that be can cause the breakdown of the program. However, when the proper attitude is displayed, the scent of the implementation of Robust Engineering, will be "sweet success."

THE CORPORATE LEADER AND THE CORPORATE TEAM

Once all levels of management have participated in a total buy-in of implementating robust engineering methodology, it is imperative that someone be chosen to be the driver, or leader of the project. This person should be a manager, highly efficient as a leader, motivated and able to motivate others. He or she should be an individual whom others would be honored to follow. He or she should be dependable and committed to the goals of the program, be firm in delegation, very detail oriented and strong in follow-up. In short, you are looking for a "champion" to "champion" the project. The organization should be certain that they have chosen the right individual for the job. At this point, an open meeting between upper management, middle management and the chosen corporate leader should take place. It is extremely important that the leader has a complete understanding of all of the elements involved, and that the tasks be broken down to be more manageable and less overwhelming. If the methodology of Robust Engineering implementation is not properly managed, it will not succeed. So, it is in the best interest of the organization that the corporate leader also be highly respected and have knowledge of the work of all departments. He or she must be able to assert authority, and be strong in project management skills. Now, the next most significant contributing factor to the success of the Robust Engineering implementation process is an empowered, multidisciplinary, cross-functional team.

Since Robust Engineering implementation is an engineering project, it would be very beneficial to involve as many engineers as possible as team members. They need to learn all of the methodology because the implementation should be considered a corporate project. A cross-functional team should consist of representatives from every department and should be a mix of personalities and talents. It is also important to note, at this point, that there should be no cultural or organizational barriers that could cause the system to break down.

All of the tasks in the implementation of Robust Engineering should be divided and shared with each team member. Each team member should take full responsibility and be held accountable for his or her part in the process. Each team member should have a clear understanding of the overall implementation process and a clear understanding of how his or her part contributes to the "big" picture. Defining the overall goals and the vision for the project will be the responsibility of a strategic planning team and the corporate leader should maintain active involvement with the strategic planning implementation process, as well as with that of the project. However, it is the responsibility of each and every member of the cross-functional team to support, acknowledge and communicate his or her meanings and objectives throughout the entire organization. For the implementation process to be successful, the structure should include control, discipline and some type of benchmarking for measurements even after it is in place and approved in order to maintain and improve the process.

Managing change is difficult in any organization, and under most circumstances. Employees have a natural resistance to and feel threatened by change. However, effective management is the key to implementing Robust Engineering, and proper project management will enhance its success. Remember, anything new requires determination and consistent encouragement, and a learning curve where mistakes and failures should be considered part of the process. It is part of the corporate leaders' job to report on completions of the different aspects of the implementation process and report the results to management. If she/he sees that something is not moving as planned, she/he must make decisions, call on her/his team and or the strategic planning team and get the process moving in the right direction. Often, when a breakdown occurs, it can be determined that communications were not clearly stated, or were misunderstood.

Effective Communication

It should be the responsibility of the cross-functional team to determine accurate and effective modes of communication. Every individual throughout the entire organization should be receiving the same messages with the exact same meanings attached to them. Robust Engineering will offer a new vocabulary, and it is the responsibility of the team to come up with interpretations that everyone will understand. The team should speculate on what some of the questions asked might be, and be prepared to answer those questions in a language which everyone will understand. Some of the phrases that may need to be expanded upon are as follows:

- What is meant by "robustness"?
- What is meant by "optimization"?
- What is meant by "variability"?
- What is "the robustness definition of failure"?
- What is the "operational meaning of prevention"?

Every employee must be approached through well-thought-out, effective communication channels. Employees will react to a message as they perceive it. The strategic planning team can put together written process information to help ensure that everyone is hearing and "getting" the same information. It is the responsibility of each cross-functional team member to see to it that their department is hearing "what was said," as well as "what was meant." Every employee needs to be able to make decisions based on the same information. Even the best of systems will fail if everyone is not pulling in the same direction. Robust Engineering methodology requires that the proper flow if information is clear, concise and timely. This is crucial to the organization. Misunderstood or misinterpreted information can deflate organizational awareness and performance.

When implementing Robust Engineering methodology, it is vital to the success of the project that lack of knowledge, misconceptions and other abnormalities be dealt with before the process is started. It is vital that employees be able to complement each others' work efforts with consistency and high performance. Each should be able to enhance the others' work by using a communication system that is not only accurate but timely, as well.

Education and Training

The importance of training and education can never be underestimated or overstated in any organization, especially when new projects or programs are being introduced into the system. Quality programs require quality initiative. A lack of understanding by even one employee can cause chaos and nonperformance. It is essential to the success of the Robust Engineering methodology process that each and every employee have an understanding of the entire process and then have a complete understanding of how his or her part in the process can contribute to the bottom line. Business literacy should be an important part of every employee's job description. Something new can be much more exciting and opportunities can be much more appreciated if employees truly feel the implications of what their contributions add. Training also helps to inspire and motivate employees to stretch their potential to achieve more. It is essential to a quality program such as Robust Engineering that everyone fully participate in all training. Engineers, especially, need to have a full understanding of the methodology and its rewards. Everyone, from top management to the lowest levels of the organization, needs to commit to focus on the goals, the needs analysis and the position of Robust Engineering, and how all these things can affect the bottom line of the organization. Knowledge adds strength to any organization, and success breeds an environment for happier co-workers.

Organizations all over the world are discovering the value of looking at their organizations from the inside out to find out what makes them work and what would make them work better. Through the development and use of an internal "expert" they are able to help their people work through changes, embrace diversity, stretch their potential and optimize their productivity. These internal experts are corporate assets. These experts must work together to come up with strategies to communicate change throughout an organization and provide accurate information. Even when quality programs already exist, the internal experts can work with the cross-functional team to help see to it that the right information is being communicated in the right ways. They are also extremely efficient and helpful in relating the goals and the purpose of Robust Engineering, as well as relating the quality principles by which the methodology is built.

THE INTEGRATION STRATEGY

Trying to implement Robust Engineering methodology and create an environment where it can be seen as part of everyone's normal work activities is a challenge in itself. There are many quality programs being used right how, and there are those who would view Robust Engineering as just another program. However, it has already been proven that Robust Engineering methodology enhances any other quality program it works with and maximizes the efficiency of the other program. Such quality programs as quality function deployment (QFD), Pugh analysis, failure modes and effects analysis

(FMEA), test planning and reliability analysis are much more effective, less time intensive, less tasteful, and give higher performance with measurable results.

Robust Engineering methodology is a brand new approach to engineering thinking. The old way of thinking was "build it, test it, fix it." The Robust Engineering way is "optimize it, confirm it , verify it." The Robust Engineering implementation process has revolutionized the engineering industry, and it's doing away with many of the more traditional tools and methods that many companies have been using. Some companies, in part, are able to change their perspectives on what constitutes failure and have been able to redefine reliability. When the emphasis is on variability, it can be associated with Robust Engineering.

When you make a product right in the first place, you don't have to make it again. Using the right materials and the right designs, saves time, saves effort, saves waste and increases the bottom line in profitability.

Bottom Line Performance

When one measures the results in any organization to pit new ways against old ways, it is usually easy to tell if one is any better than the other. But when one measures and can find considerable differences, a closer look is certainly merited. The implementation of Robust Engineering will come out ahead on the bottom line, any day of the week. What it really says is, Would you rather have your engineers spending their days roaming the organization to put out fires, or do you think you would foster better performance by having them prevent the fires by preventing the causes? When you can link quality, cost and time, to real success, you can usually trace it to bottom-line performance. And in addition to revealing your bottom-line success, what if you offered incentives to your workers, the ones who actually made it all happen? What if you dangled in from of them percentages of bottom-line dollar differences? Your employees will feel very important to the organization when they can actually see how they have helped to enhance the bottom lines, but they will feel sheer ecstacy when they are able to share in it.

CHAPTER THREE 

Robust Engineering Process Formula

A CASE STUDY: OPTIMIZATION OF INTER-COOLER BY NISSAN MOTOR COMPANY

INTRODUCTION

In this chapter a case study will be discussed to illustrate the thought process and steps utilized to formulate and execute a Robust Engineering project. Nissan Motor Company conducted the case study on an inter-cooler in 1996.

The steps to formulate a Robust Engineering optimization are as follows.

Plan

Step 1: Define the scope of project.
Step 2: Define the boundary of subsystem.
Step 3: Define the input signal M and the output response y, and the ideal function.
Step 4: Develop signal and noise strategies.
Step 5: Define control factors and levels.
Step 6: Formulate the experiment and prepare for the experiment.

Do

Step 7: Conduct the experiment/simulation and collect data.
Step 8: Conduct data analysis.
Step 8-1: Calculate signal-to-noise ratio and sensitivity for each run.
Step 8-2: Generate a response table for S/N and sensitivity and study and interpret the response tables.
Step 8-3: Conduct two-step optimization.
Step 8-4: Make predictions.

Check

Step 9: Conduct confirmation run and evaluate the reproducibility.

Act

Step 10: Document and implement the result.
Step 11: Plan the next step.

Inter-Cooler Robust Design Optimization, Nissan Motor Company

Project Background

Automotive customers are becoming more and more sensitive to vehicle audible noises. The audible noise problem caused by the airflow through the inter-cooler used for a supercharger has become a critical issue. Traditional approaches to reduce audible noise have typically been ineffective and inefficient, and more important have resulted in a cost increase.

In this study, the robust design approach using ideal function was applied to reduce variability of the airflow through the inter-cooler. As a result, the optimized parameter setting not only achieved a significant reduction in audible noise, but also improved the cooling efficiency and reduced the cost by more than $3.50 per unit.

The System

One of the means to enhance the engine power is composed of a turbocharger (hereinafter referred to as T/C) and an inter-cooler (I/C). The system of T/C and I/C is shown in Figure 1. Engine power depends on the quantity of air-fuel mixture charged into an engine cylinder. The more the air-fuel mixture is charged, the more the engine power increases.

In contrast to an ordinary natural suction engine, T/C compresses the air well above the atmospheric pressure. A turbine driven by exhaust gas drives a compressor to compress the airflow. Compressed air into a cylinder drastically enhances the air-fuel mixture charging efficiency.

While compressed air improves the efficiency in charging when air is compressed, the air temperature rises and a higher temperature results in decreased air density. Reduction of air density results in a decreased charging

Figure 1: The System – T/C and I/C

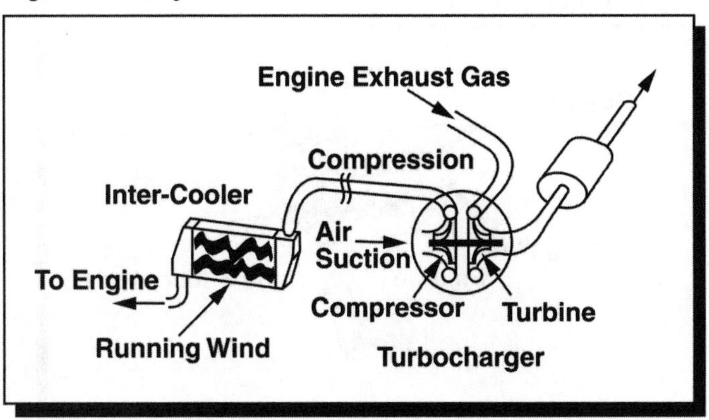

efficiency. In order to resolve this conflict, I/C is added to cool the compressed air generated by T/C. Figure 1 shows T/C and I/C. In other words, T/C and I/C together will result in delivering low temperature and highly compressed air into a cylinder for combustion.

Define the Scope of the Project and the Boundary of the Subsystem (Steps 1 and 2)

Step 1: Define the scope of the project and Step 2: Define the boundary of the subsystem are probably the most important steps. It is extremely important to focus on the energy transformation and not to focus on symptoms of the variability in energy transformation. The focus also depends on criteria such as:

- Scale of System and Subsystems: Usually, it is better to focus on subsystems rather than a big system. It depends on the number of subsystems in the system, the interaction between the subsystems and the number of design parameters in the subsystem.
- Measurement Technology: This depends on the feasibility/cost of measurement technology. Sometimes one may have to develop a fixture in order to measure the energy transformation.
- Computer Simulation: This depends on whether the function can be simulated by a computer and on what can be simulated. Usually, it is much cheaper to run the experiment by computer simulation than by hardware.

The Nissan study used the following thought process.

- The I/C schematic is shown in Figure 2.
- Compressed high temperature air from T/C is cooled through the tube with fins in the I/C. As the air flows through the tube, its resistance is high, which leads to poor cooling performance and causes low air-charging efficiency.

Figure 2: Construction of I/C

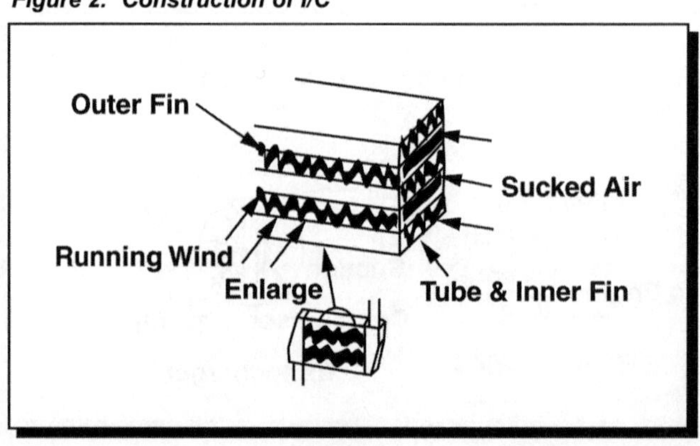

- Another issue is that when airflow is not uniform within the I/C, it causes not only poor air-cooling performance but also audible noises.
- In traditional product development, whenever a problem arises, we study the cause of the problem and seek a solution for each problem. In the case of I/C design, we typically study characteristics such as heat radiation, pressure drop, airflow audible noise (sound pressure), cooling temperature, heat-exchanging efficiency and engine power. These characteristics are measured because they are in the list of requirements. Then whenever the requirement is not met, we seek a solution.
- Previous to this study, Nissan could not solve the problem of audible noise. They designed an experiment with sound pressure (air flow audible noise level) as the response in an attempt to minimize the response. It was not successful.
- The scope of this project is to optimize the I/C design for its robustness based on energy transformation.

Ideal Function, Input Signal and Output Response (Step 3)

This step is used to define the input signal M and the output response y. It is based on the energy transformation of the system. (Sometimes we have to search for M and y, in that they represent the energy transformation, when measuring the energy itself is not feasible.) Then the ideal function must be defined. The ideal function is the ideal relation between M and y, based on the physics of the system.

The function of I/C is to cool the high temperature airflow compressed by T/C. This heat-exchanging function can be formulated as I/C's ideal function. It can be expressed as:

Ideal Function

Ideal Function:	$y = \beta M$	
where	Input Signal:	M = Temperature Differential
	Output Response:	y = Temperature Reduction

Another approach is to focus on the airflow generation in I/C; the logic is that if airflow in I/C is flowing smoothly and efficiently, heat exchanging also will be efficient.

In this study, the airflow generation aspect was taken for the ideal function rather than the heat-exchanging aspect due to the ease of measurement. The ideal function based on physics is illustrated below and in Figure 3.

Ideal Function

Ideal Function:	$y = \beta M$	
where	Input Signal: Output Response:	M = Theoretical Airflow thru I/C y = Actual Airflow thru I/C
Note:	For each run of experiment, T/C RPM and Input airflow were varied experimentally, then M is calculated by a formula based on physics. M = f (T/C RPM, Input Airflow, Tube Cross-Sectional Area)	

Figure 3: Ideal Function of I/C

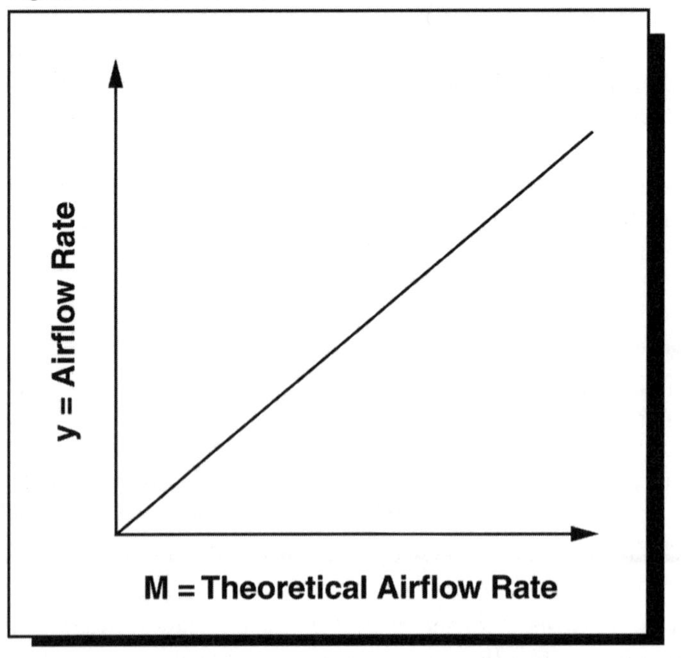

Develop Signal and Noise Strategies (Step 4)

In a robust design optimization, the signal factor M will be varied intentionally. It will be varied to cover the customer's usage conditions. In this study, the signal factor M is defined as theoretical airflow. In order to generate different values of M, the input airflow into T/C and T/C RPM were set and varied to cover the range of operation. T/C RPM was varied from 60,000 to 100,000 RPM, and input airflow from 6.0 to 10.0 kg/min. As a result, seven levels of M were defined as shown in the table below.

Signal-to-Noise

	M1	M2	M3	M4	M5	M6	M7
T/C RPM (x 10^4)	6.0	8.0	8.0	8.0	10.0	10.0	10.0
Input Airflow (kg/min)	6.0	6.0	8.0	10.0	6.0	8.0	10.0

Then the value of M will be calculated as a function of T/C RPM, input airflow and cross-sectional area. Cross-sectional area is determined by the combination of control factors.

In a robust design optimization, it is extremely important to measure the response under more than one noise condition. There are three major categories of noises:

Environment: Customers' usage conditions
Aging: Aging, wear and deterioration
Manufacturing: Manufacturing variation in material and dimensions

It is most efficient to apply "compounded noise strategy." A compounded noise strategy generates two noise conditions as follows:

N1 = Noise condition such that the energy transformation tends to be on the low side.

N2 = Noise condition such that the energy transformation tends to be on the high side.

N1 and N2 are generated by mixing (compounding) a few noise factors from the customer usage environment and aging. It is extremely important to optimize for robustness against noises of environment and aging.

In the I/C robust design study, noise factors from the environment and aging were not taken into consideration. (Mediocre noise strategy is the only weakness of this study.) Figure 4 shows the details of I/C and two levels of noise factor, namely, I = Position within I/C. Since the variability of airflow within I/C tubes was so large, the noise factor was taken only from position to position and within position.

Noise Factors and Levels

		Level 1	Level 2
I	Position within I/C	Upper Section	Lower Section
J	Airflow within the Position	Maximum Flow	Minimum Flow

Figure 4: Noise Factors and Levels

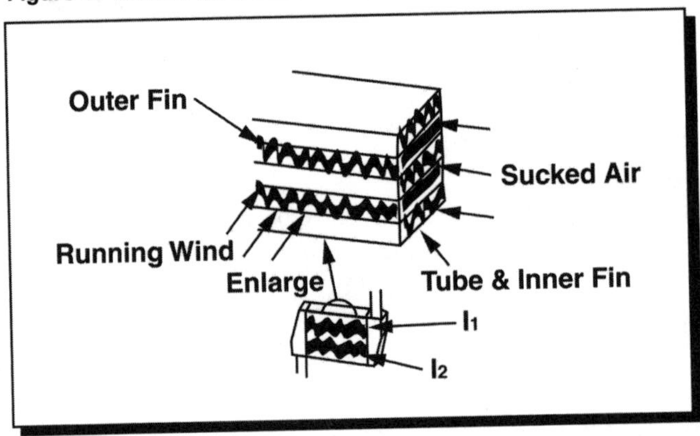

The performance of initial design with respect to the ideal function and the noise effects are shown in Figure 5.

Figure 5: The Performance of Initial Design

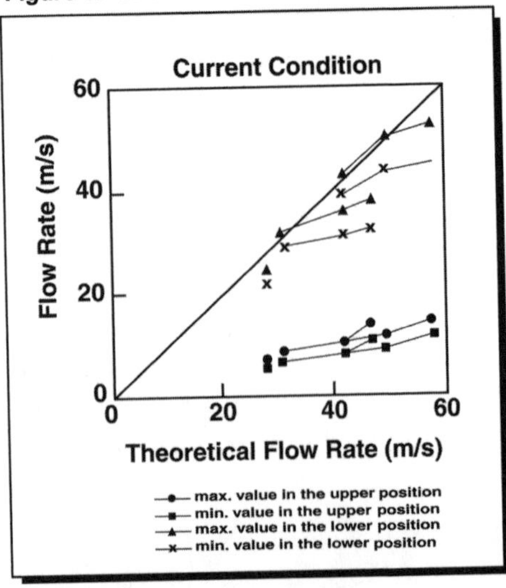

Define Control Factors and Levels (Step 5)

In order to achieve a robust function, we need to try out various design alternatives. Control factors (design parameters) and their levels (alternatives) are selected for robustness evaluation. Within the design space, we want to try various design alternatives to search for the most robust design. Constraints such as cost/weight/size must be taken under consideration when control factors and levels are selected. (Use engineering common sense.)

Control factors and levels, noise factors and levels and signal factors and levels are summarized in Table 1.

Table 1: Factors and Levels for Analysis

Factor		Level	1	2	3
Control Factor	A	Length of Inlet Tank (mm)	Standard *	Standard +20	----
	B	Thickness of Tube (mm)	5 *	7	9
	C	Shape of Inlet Tank	Shape 1	Shape 2 *	Shape 3
	D	Flow Direction of Inlet Tank	Direction 1 *	Direction 2	Direction 3
	E	Length of Tube (mm)	120	140 *	160
	F	Shape of Tube end (deg.)	0	30 *	60
	G	Length of Inner Fin (mm)	Standard –7.5	Standard *	Standard +7.5
	H	Inner Dia. of Inlet Tube (mm)	ø 45	ø 55 *	ø 65
Noise Factor	I	Position within I/C	Upper	Lower	----
	J	Airflow within Position	Max. Flow	Min. Flow	----
Signal Factor	M	Theoretical Flow Rate (m/s)	Level on operating condition (combination of T/C revolution and airflow)		

* *Indicates a current level.*

FORMULATE AND PREPARE FOR THE EXPERIMENT, CONDUCT THE EXPERIMENT AND COLLECT DATA (STEPS 6 AND 7)

Control factors A to H (one two-level factor and seven three-level factors) were assigned to a standard orthogonal array L_{18}, as shown in Table 2. An orthogonal array provides an orthogonal design of the experiment matrix. By assigning factors to the columns of an orthogonal array, the experimentation runs will be balanced. This property is very powerful compared to a traditional shotgun approach or a one-factor-at-a-time approach. When an L_{18} is used there will be 18 combinations (18 recipes) of control factors or 18 experimental runs. In other words, 18 I/C design configurations will be tried out for data collection.

To illustrate the use of an L_{18}, for instance, experiment No. 1 will be an A1, B1, C1, D1, E1, F1, G1, H1 configuration with an I/C design having the standard length of the inlet tank 5-mm tube thickness, shape 1 of inlet tank, flow direction-1, 120-mm tube length, 0 degree tube end, inner fin length of standard –7.5 mm and 45-mm inner diameter. Recognize that there would be 18 recipes of I/C design altogether for this experiment.

Table 2: Control Factor Layout with L_{18}

Run No.	A 1	B 2	C 3	D 4	E 5	F 6	G 7	H 8	A: Length	B: Thickness	C: Shape	D: Direction	E: Tube Length	F: End Shape	G: Fin Length	H: Diameter
1	1	1	1	1	1	1	1	1	Std	5	1	1	120	0	–7.5	45
2	1	1	2	2	2	2	2	2	Std	5	2	2	140	30	Std	55
3	1	1	3	3	3	3	3	3	Std	5	3	3	160	60	+7.5	65
4	1	2	1	1	2	2	3	3	Std	7	1	1	140	30	+7.5	65
5	1	2	2	2	3	3	1	1	Std	7	2	2	160	60	–7.5	45
6	1	2	3	3	1	1	2	2	Std	7	3	3	120	0	Std	55
7	1	3	1	2	1	3	2	3	Std	9	1	2	120	60	Std	65
8	1	3	2	3	2	1	3	1	Std	9	2	3	140	0	+7.5	45
9	1	3	3	1	3	2	1	2	Std	9	3	1	160	30	–7.5	55
10	2	1	1	3	3	2	2	1	+20	5	1	3	160	30	Std	45
11	2	1	2	1	1	3	3	2	+20	5	2	1	120	60	+7.5	55
12	2	1	3	2	2	1	1	3	+20	5	3	2	140	0	–7.5	65
13	2	2	1	2	3	1	3	2	+20	7	1	2	160	0	+7.5	55
14	2	2	2	3	1	2	1	3	+20	7	2	3	120	30	–7.5	65
15	2	2	3	1	2	3	2	1	+20	7	3	1	140	60	Std	45
16	2	3	1	3	2	3	1	2	+20	9	1	3	140	60	–7.5	55
17	2	3	2	1	3	1	2	3	+20	9	2	1	160	0	Std	65
18	2	3	3	2	1	2	3	1	+20	9	3	2	120	30	+7.5	45

(Yes, conducting these 18 experimental runs are a lot of work. It seems time consuming and costly. But the payback often far exceeds the cost of the experiment, say 10 to 100 times. It really requires management support to run this type of experimentation.)

Eighteen I/Cs were built and the response y = airflow was measured. For illustration, the data set for run No. 1 of L_{18} is shown in Table 3. The data set for No. 1 was obtained as follows. For each of seven combinations of T/C RPM and input airflow, the airflow rate through the I/C tubing was measured at the upper and lower sections. When the T/C RPM was set at 60,000 RPM and the input airflow at 6.0 kg/min, four airflow rate measurements were recorded as data, namely 3.4, 2.7, 15.4 and 13.6. These four data represent the maximum and minimum flow rates at the lower and upper positions. Then T/C RPM was set at 80,000 RPM and input airflow at 6.0 kg/min, and the measured airflow rates were 5.6, 4.4, 21.8 and 18.3, respectively.

The values of M, namely 29.0, 30.1, 40.0 · · ·, are computed as a function of T/C RPM, input airflow and cross-sectional area. Cross-sectional area is determined by the particular combination of control factor levels, therefore it will be different for each of 18 configurations. Thus, M must be calculated for each run of L_{18}.

In most robust design optimization studies, the value of M is set by experiment and it is not necessary to compute each run as was done in this case study. Recognize that there would be a set of data like those in Table 3 for each of 18 configurations of L_{18}.

Table 3: Data Set for Run No. 1 of L_{18}

T/C RPM (x 10^4 RPM)	6.0	8.0	8.0	8.0	10.0	10.0	10.0
Input Airflow (kg/min)	6.0	6.0	8.0	10.0	6.0	8.0	10.0
M: Theoretical Airflow Rate	**29.0**	**30.1**	**40.0**	**46.1**	**41.3**	**49.5**	**58.1**
I_1,J_1 = Upper Position - Max. Flow	3.4	5.6	6.7	9.0	8.5	10.7	12.2
I_1,J_2 = Upper Position - Min. Flow	2.7	4.4	5.4	7.0	6.5	8.4	9.8
I_2,J_1 = Lower Position - Max. Flow	15.4	21.8	24.0	26.3	31.3	36.0	41.7
I_2,J_2 = Lower Position - Min. Flow	13.6	18.3	20.0	21.8	26.8	30.0	34.4

Figure 6: Response From Run No. 1 of L_{18}

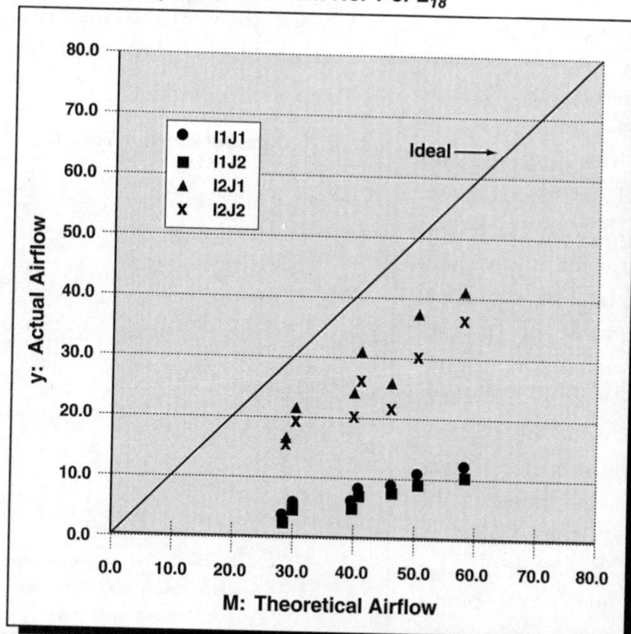

Figure 6 shows the response y = actual airflow plotted against M = theoretical (ideal) airflow for run No. 1 of L_{18}. Notice that the performance is not good. Ideally, all responses must be on the 45-degree line, which would indicate that the I/C is producing the airflow with 100% efficiency and no variability.

Conduct Data Analysis (Step 8)

Calculate Signal-to-Noise Ratio and Sensitivity for Each Run

Signal-to-noise ratio and sensitivity are calculated for each run of L_{18}. The concept of the S/N ratio and sensitivity are illustrated using the data set from No. 1 of L_{18}. The S/N ratio and sensitivity are given by the following:

$$S/N = \eta = 10 \log \frac{\beta^2}{\sigma^2}$$

$$S = \text{Sensitivity (Efficiency)} = 10 \log \beta^2$$

Figure 7 shows the data set from run No. 1 of L_{18}. The diagonal line through the zero point represents the so-called least square best-fit line through all data. This line is forced through the zero point. The slope of this line is denoted by β and it is called sensitivity, i.e., sensitivity of y with respect to the input signal M. Recognize that β also represents the efficiency of airflow generation by I/C.

What if β is zero? This would indicate no response or the that the I/C does not generate airflow, i.e., that the I/C does not work. Notice that β^2 is the output power the I/C is supposed to generate. It is indeed proportional to the useful energy the I/C is delivering.

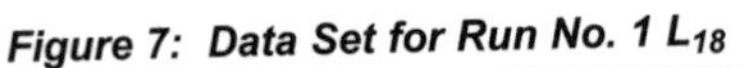
Figure 7: Data Set for Run No. 1 L_{18}

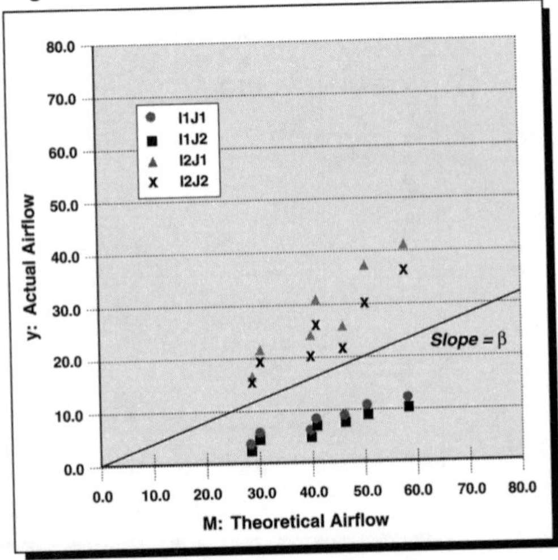

σ^2 is called the mean square around the best-fit line. It is the average of squares of distances from an individual point to the best-fit line. The smaller the value of σ^2, the smaller the variability around the best-fit line is. Or the smaller the value of σ^2, the smaller the noise effects are. Notice that the higher S/N indicates that the system is efficiently producing the response without being affected by noises. σ^2 is proportional to work generated by noise factors. The S/N ratio indeed is the measure of robustness.

The equations to calculate the S/N ratio and sensitivity are demonstrated using the No. 1 data set as follows. Note that S/N is denoted by η.

$$S_T = 3.4^2 + 5.6^2 + 6.7^2 + \cdots + 34.4^2 = 11{,}009.41$$
$$r = 29.0^2 + 30.1^2 + 40.0^2 + 46.1^2 + 41.3^2 + 49.5^2 = 26{,}007.54$$
$$L_1 = 29.0 \times 3.4 + 30.1 \times 5.6 + 40.0 \times 6.7 + \cdots + 58.1 \times 12.2 = 2{,}539.58$$
$$L_2 = 29.0 \times 2.7 + 30.1 \times 4.4 + 40.0 \times 5.4 + \cdots + 58.1 \times 9.8 = 2{,}003.07$$
$$L_3 = 29.0 \times 15.4 + 30.1 \times 21.8 + 40.0 \times 24.0 + \cdots + 58.1 \times 41.7 = 8{,}772.67$$
$$L_4 = 29.0 \times 13.6 + 30.1 \times 18.3 + 40.0 \times 20.0 + \cdots + 58.1 \times 34.4 = 7{,}340.69$$

$$S_\beta = \frac{(L_1 + L_2 + L_3 + L_4)_2}{4 \times r} = 8{,}202.828 \qquad (f=1)$$

$$S_{\beta \times N} = \frac{(L_1 + L_2 + L_3 + L_4)_2}{r} - S_\beta = 8{,}202.828 \qquad (f=3)$$

$$S_e = S_T - S_\beta - S_{\beta \times N} = 142.775$$
$$V_e = S_e / 24 = 5.9490$$
$$V_N = (S_T - S_\beta) / 24 = 103.9475$$

$$\eta = 10 \log \frac{\frac{1}{4 \times r}(S_\beta - V_e)}{V_N} = -28.19 \text{ dB}$$

$$S = 10 \log \frac{1}{4 \times r}(S_\beta - V_e) = -8.02 \text{ dB}$$

Table 4: The S/N Ratio and Sensitivity

No.	A	B	C	D	E	F	G	H	S/N Ratio η	Sensitivity S
1	1	1	1	1	1	1	1	1	-28.19	-8.02
2	1	1	2	2	2	2	2	2	-19.11	-5.72
3	1	1	3	3	3	3	3	3	-21.52	-5.59
4	1	2	1	1	2	2	3	3	-27.32	-2.83
5	1	2	2	2	3	3	1	1	-18.95	-0.59
6	1	2	3	3	1	1	2	2	-12.76	-1.51
7	1	3	1	2	1	3	2	3	-17.08	-5.61
8	1	3	2	3	2	1	3	1	-22.65	0.41
9	1	3	3	1	3	2	1	2	-28.33	-2.19
10	2	1	1	3	3	2	2	1	-17.71	-5.54
11	2	1	2	1	1	3	3	2	-29.04	-6.79
12	2	1	3	2	2	1	1	3	-22.51	-4.35
13	2	2	1	2	3	1	3	2	-23.24	-4.55
14	2	2	2	3	1	2	1	3	-21.89	-4.75
15	2	2	3	1	2	3	2	1	-29.16	-6.11
16	2	3	1	3	2	3	1	2	-24.89	-3.61
17	2	3	2	1	3	1	2	3	-25.83	-4.49
18	2	3	3	2	1	2	3	1	-17.50	-2.98

For each of 18 sets of data, this calculation is repeated. The results are shown in Table 4.

Generate Response Tables and Graphs for the S/N Ratio and Sensitivity

Response tables are generated for both the S/N ratio and sensitivity. The results are shown in Table 5 and Figure 8. The response table shows the average S/N ratio and sensitivity for each factor level.

For example, the average S/N for A_1 is calculated by averaging S/N values for No. 1 through No. 9, where the level of factor A is at level 1. Likewise, the average of A_2 is calculated by averaging S/N from No. 10 through No. 18. Similar calculations are conducted for B through H. The response table for sensitivity is calculated in a similar fashion.

Sample Calculation of S/N Response

$$\overline{A}_1 = \frac{-28.19 - 19.11 - 21.52 - 27.32 - 18.95 - 12.76 - 17.08 - 22.65 - 28.33}{9} = -21.77$$

$$\overline{A}_2 = \frac{-17.71 - 29.04 - 22.51 - 23.24 - 21.89 - 29.16 - 24.89 - 25.83 - 17.50}{9} = -23.53$$

$$\overline{B}_1 = \frac{-28.19 - 19.11 - 21.52 - 17.71 - 29.04 - 22.51}{6} = -23.01$$

$$\overline{B}_2 = \frac{-27.32 - 18.95 - 12.76 - 23.24 - 21.89 - 29.16}{6} = -22.22$$

$$\overline{B}_3 = \frac{-17.08 - 22.65 - 28.33 - 24.89 - 25.83 - 17.50}{6} = -22.71$$

Figure 8: S/N Ratio and Sensitivity

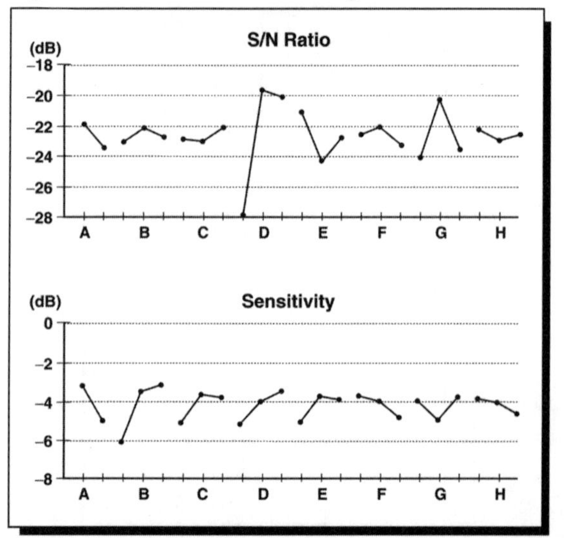

Table 5: Response Table for S/N Ratio and Sensitivity

Control Factor \ Feature		S/N Ratio (dB) 1	S/N Ratio (dB) 2	S/N Ratio (dB) 3	Sensitivity (dB) 1	Sensitivity (dB) 2	Sensitivity (dB) 3
A	Length of Inlet Tank (mm)	–21.77	–23.53	----	–3.52	–4.79	----
B	Thickness of Tube (mm)	–23.01	–22.22	–22.71	–6.01	–3.39	–3.07
C	Shape of Inlet Tank	–23.07	–22.91	–21.96	–5.03	–3.65	–3.78
D	Flow Direction of Inlet Tank	–27.98	–19.73	–20.23	–5.07	–3.96	–3.43
E	Length of Tube (mm)	–21.08	–24.27	–22.60	–4.94	–3.70	–3.83
F	Shape of Tube End (deg.)	–22.53	–21.98	–23.44	–3.76	–3.99	–4.72
G	Length of Inner Fin (mm)	–24.13	–20.27	–23.54	–3.92	–4.83	–3.71
H	Inner Diameter of Inlet Tube (mm)	–22.36	–22.89	–22.69	–3.80	–4.06	–4.60
	EXPERIMENTAL AVERAGE VALUE		–22.65			–4.15	

Conduct Two-Step Optimization

Two-step optimization is as follows:

Step 1: Reduce functional variability (maximize S/N ratio)
Step 2: Adjust β

Step 1 is used to maximize the S/N ratio. Then step 2 is employed to adjust sensitivity, i.e., β. For an inter-cooler, a high sensitivity is desired for a high efficiency. Nissan selected the optimum design by giving the highest priority to maximizing the S/N ratio and cost reduction. For example, B_1 is selected as the optimum level since B_2 and B_3 will require a drastic change in the fabrication process. Likewise, D_3 is selected for having a great advantage in vehicle assembly. F_1, G_1 and H_2 are selected to reduce cost without sacrificing the S/N ratio unduly. As compared to the initial design factors C, D and E have contributed greatly to reducing variability of airflow rate.

	A	B	C	D	E	F	G	H
Initial Design	1	1	2	1	2	2	2	2
To Maximize The S/N Ratio	1	2	3	2,3	1	1,2	2	1
Optimum Design	1	1	3	3	1	1	1	2

Make Predictions

Predictions on performance under optimum design and initial design can be made. The predictions are made using the additivity of factorial effects. The results are shown in Table 6.

Table 6: Prediction and Confirmation Results

	Prediction		Confirmation	
	S/N Ratio	Sensitivity	S/N	Sensitivity
Initial Design	–26.55	–5.76	–28.51	–5.15
Optimum Design	–19.07	–4.33	–19.74	–4.22
Gain	**7.48**	**1.43**	**8.77**	**0.93**

The prediction was to improve the S/N ratio by 7.48 dB and sensitivity by 1.43 dB. Since a 6-dB gain in S/N is equivalent to halving the range of variability, a gain of 7.48 dB in S/N implies a variability reduction of 58%. A gain of 1.43 dB in sensitivity implies an 18% increase in airflow generation efficiency. Now, all these are only predictions. We must confirm the results. It is not yet Miller time.

Conduct A Confirmation Run (Step 9)

It is absolutely necessary to conduct a confirmation run. Figure 9 shows the results from the confirmation run. The S/N ratio and sensitivity are calculated from the confirmation data, as shown in Table 6.

It shows very good reproducibility. The predicted improvement is fairly close to the actual result from the confirmation run.

An 8.77-dB gain in the S/N ratio is equivalent to reducing the variability range by 64%. The sensitivity gain was 0.93 dB in the confirmation. This increases the slope by 11%; the efficiency in airflow rate was improved by 11%.

Figure 9: Confirmed Data on Flow Rate

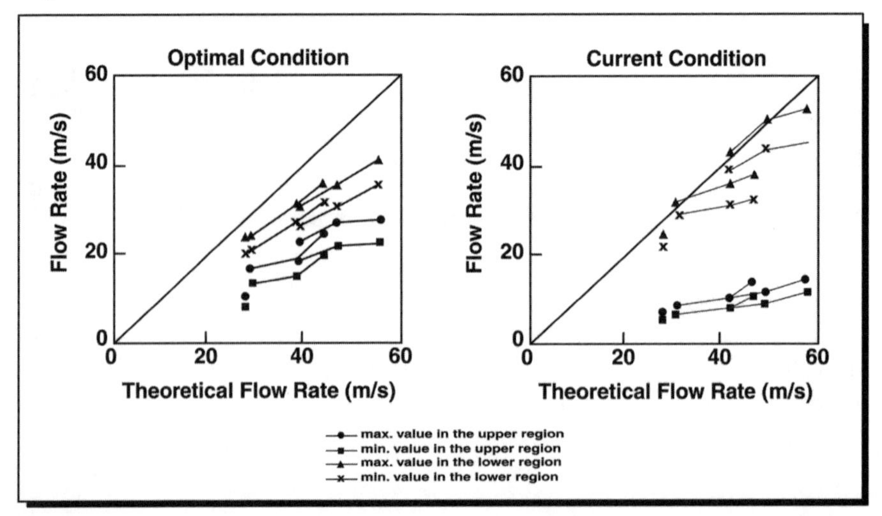

Nissan has confirmed great improvement in those traditional responses/requirements. Figure 10 show the reduction in "audible noise." The sound pressure at the noisiest operation condition has become 6 dBA quieter than in the initial design. It is a tremendous improvement in audible noise reduction.

Figure 10: Confirmed Results on Airflow Noise

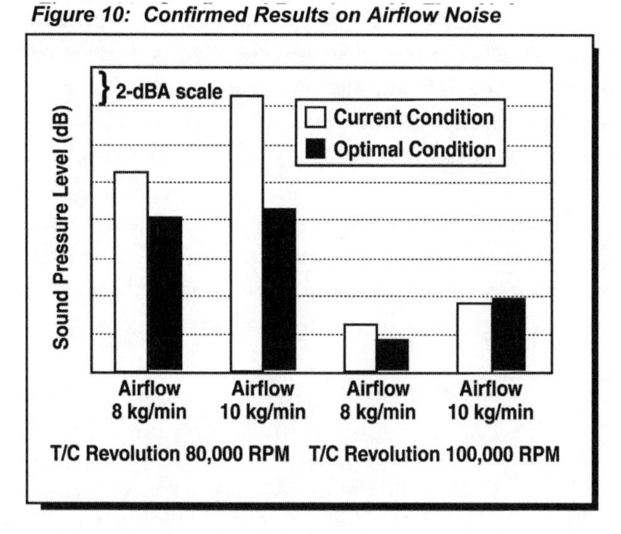

Figure 11 shows an improvement in "cooling performance." As a result of optimization, 20% improvement in the cooling performance was confirmed. It is believed that both "variability reduction" and "increase in airflow efficiency" contributed to this drastic improvement in cooling performance.

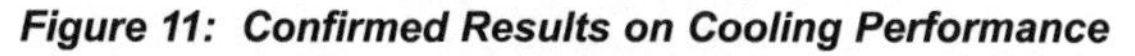

Figure 11: Confirmed Results on Cooling Performance

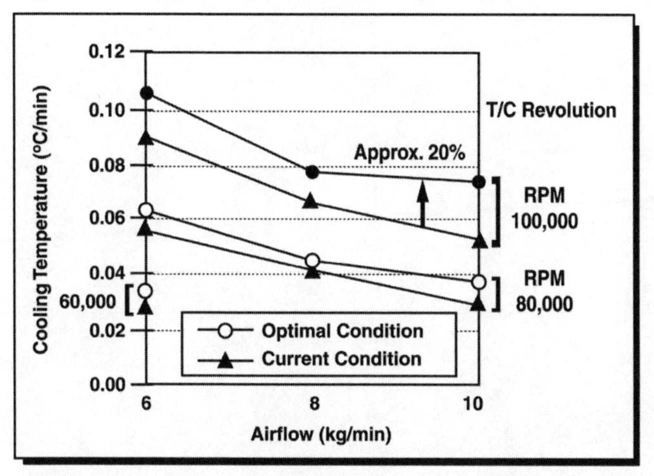

This implies that to achieve the same cooling capacity for an I/C, the I/C can be sized down by 20%, leading to not only cost reduction but also to a more flexible layout in the engine room during vehicle development. Nissan did not give out the exact cost reduction figure; however, the author was told it was at least $3.50 per I/C.

As an aside, Nissan said they conducted the same L_{18} experiment by measuring the "sound pressure" as a response and the results did not confirm this. They could not reduce the I/C audible noise, and were left highly frustrated concerning use of the design of the experiment approach. Luckily, they did not scrap those 18 configurations (prototypes) used for that DoE.

Then they used those same samples, but measured the energy transformation response introduced in this study and confirmed great improvements in all aspects of I/C performance. Now they have a quiet, highly efficient and flexible inter-cooler design for a family of products and future products.

CHAPTER FOUR

Accuracy Improvement of a Disposable Oxygen Sensor Used for Open Heart Surgery

CDI/3M HEALTH CARE – USA

EXECUTIVE SUMMARY

BACKGROUND

3M Healthcare uses a technology called optical fluorescence microsensing to produce a line of products that monitor patients' blood gas levels in the extracorporeal circuit during bypass surgery. There are several steps involved in the blood gas monitoring system. It became the need of the hour to optimize the design of the small disk of polymeric substrate, called a sensor, for a consistent and predictable emission.

PROCESS

A Taguchi double-signal optimization technique was used to improve the accuracy of a disposable blood gas sensor that is used during open-heart surgery. Factors optimized were mainly chemical in nature, as the function of the sensor is based on a chemical fluorescence principle. In this study seven control factors and three noise factors are used for optimization. Actual oxygen concentration and temperature of blood were considered as input signals.

True oxygen concentration and blood temperature were taken as signal factors in the measured value of oxygen concentration. Blood temperature is a signal because blood temperature varies from operation to operation from 18°C to 37°C, and, hence, there is a need to compensate for the effect of blood temperature. Seven design parameters are optimized for robustness against selected noise conditions.

BENEFITS

- This study yielded a sensor design that emits a very consistent and predictable fluorescence based on the patient's blood oxygen concentration and temperature, the two signal factors in the study.

- A threefold improvement in 9 dB in S/N ratio accuracy was realized in comparison to the best previous design. This translates to a reduction in measurement errors by more than 66%.
- The study led to a few important findings including how the effect of temperature can be compensated.
- The new design was optimized for robustness against the noise factors.

CASE STUDY

INTRODUCTION

3M Healthcare uses a technology called optical fluorescence microsensing to produce a line of products that monitor patients' blood gas levels in the extracorporeal circuit during bypass surgery. The extracorporeal circuit, as Figure 1 depicts, transfers blood from the patient to pumps where it is oxygenated and replaces the function of the heart and lungs during the surgical procedure; it then returns the oxygenated blood back to the patient. One of the functions of our monitoring system is to give the operating room team a continuous readout of the patient's oxygen concentration (measured in units of partial pressure, mmHg), both before and after use of the oxygenating equipment. The before or "venous" measurement is used by the team to assess the patient's physiological and metabolic state, while the after or "arterial" measurement is used as feedback so that the oxygenation equipment can be adjusted by the "perfusionist" manning the extracorporeal station.

Figure 1: CDI Monitoring System Installed in the Extracorporeal Circuit During Bypass Surgery

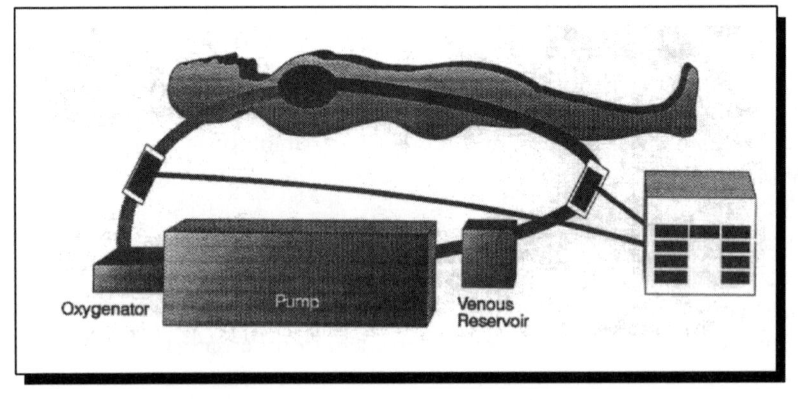

The blood gas monitoring systems operate as follows (see Figure 2):

1. Oxygen from the patient's blood diffuses through a sterile membrane and into a small disk of polymeric material containing a fluorescing dye (this disk is called the SENSOR).
2. Light of a specific wavelength is generated and delivered by a fiber-optics cable to irradiate the sensor.

3. Once excited, the sensor chemistry fluoresces depending on the concentration of oxygen dissolved in the sensor.
4. Another fiber-optics bundle transfers a portion of the resulting fluorescence to a photodetector that translates the light signal to an analog signal proportional to the intensity of light received.
5. Further analog and digital processing converts the raw fluorescence to a partial pressure displayed in mmHg, the measure of oxygen concentration needed.

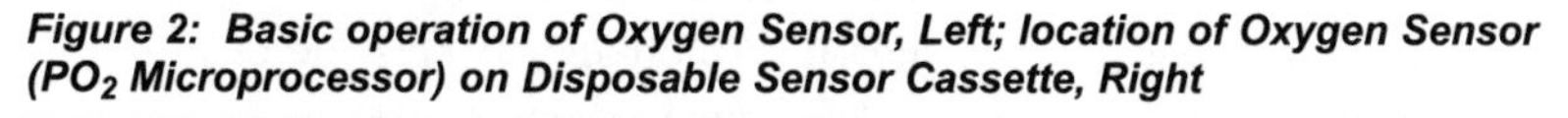

Figure 2: Basic operation of Oxygen Sensor, Left; location of Oxygen Sensor (PO_2 Microprocessor) on Disposable Sensor Cassette, Right

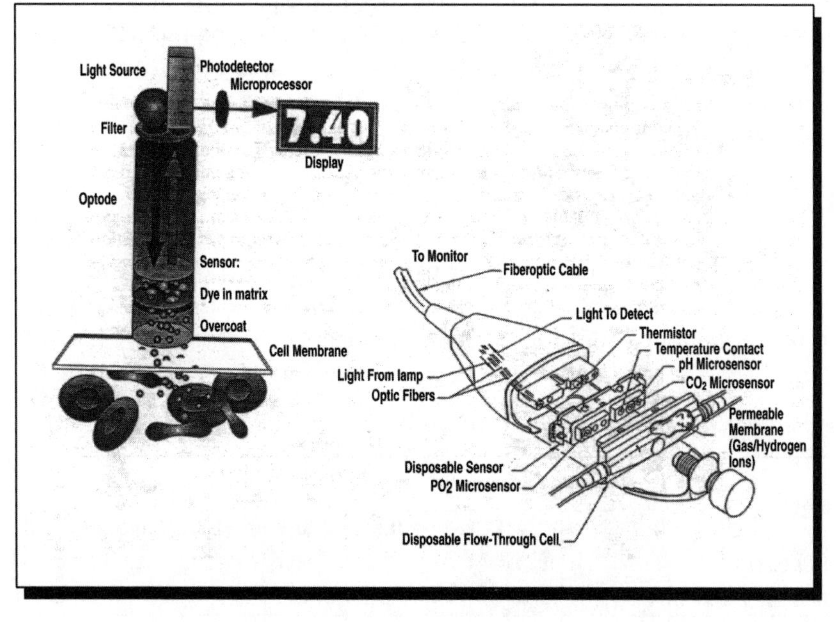

The objective of this study was to optimize the design of the small disk of polymeric chemistry which we call the SENSOR so that it emits a very consistent and predictable fluorescence based on the patient's blood oxygen concentration and temperature, the two signal factors in the study.

Taguchi Methods were the preferred approach to optimize the design of the sensor because they offer an efficient way to make a design robust to noises, which we will surely encounter in future manufacturing and product use. The "dynamic" optimization approach was preferred over other methods because we wanted to improve the predictability of the sensor over the full range in which it operates: oxygen concentrations of 20 to 300 mmHg and temperatures of 18 to 37° C. It was decided to use the "double-signal' technique for optimizing the sensor's design because the output of a sensor (fluorescence) should depend on two things: (1) the actual oxygen concentration of the blood (primary signal), and (2) the temperature of the blood (secondary signal). Since there are two inputs, or signals, to the sensor, this was a perfect application for the double-signal approach.

Although it's easy to understand why the actual blood oxygen concentration is an input or signal to the sensor, it's not so easy to understand why temperature is an input. The reason it is a valid input is that the temperature of the blood affects the solubility of oxygen in the sensor, just like temperature affects the solubility of most gases in liquids. Since temperature affects how much oxygen can dissolve in the sensor, it then also affects the resulting fluorescence, or output, of the sensor. We have in the past used an algorithm in our monitor's software to do a feed forward correction of the oxygen readout for temperature. One of the objectives of this study was to improve the accuracy of this coefficient by modifying the design of the sensor so that its temperature effect was more predictable and consistent.

Experimental Methods – Factors and Design Format

The factors chosen to be optimized in this experiment were thought to be key design factors influencing the performance of the sensor disk. Factors were segregated depending on whether they were control, noise or signal factors and the optimization was done using an L_{18} orthogonal array (see Figure 3).

Figure 3: Experimental Design Format

Experimental Design Format

							M1	M1	M1	M1	M1	M1	M2	M2	M2	M2	M2	M2	M3	M3	M3	M3	M3	M3	M4	M4	M4	M4	M4	M4
MODIFIED L18							T1	T2	T3	T1	T2	T3	T1	T2	T3	T1	T2	T3	T1	T2	T3	T1	T2	T3	T1	T2	T3	T1	T2	T3
OT	t	R	OV	S	MW	C	N1	N1	N1	N2	N2	N2	N1	N1	N1	N2	N2	N2	N1	N1	N1	N2	N2	N2	N1	N1	N1	N2	N2	N2
1	1	1	1	1	1	1																								
1	2	2	2	2	2	2																								
1	3	3	2	3	3	3																								
2	1	1	2	2	3	3																								
2	2	2	2	3	1	1																								
2	3	3	1	1	2	2																								
3	1	2	1	3	2	3																								
3	2	3	2	1	3	1																								
3	3	1	2	2	1	2																								
4	1	3	2	2	2	1																								
4	2	1	1	3	3	2																								
4	3	2	2	1	1	3																								
5	1	2	2	1	3	2																								
5	2	3	1	2	1	3																								
5	3	1	2	3	2	1																								
1	1	3	2	3	1	2																								
1	2	1	2	1	2	3																								
1	3	2	1	2	3	1																								

OT = Overcoat Type
t = Sensor Thickness
R = Dye/Polymer 1 Ratio
OV = Overcoat Parameter
S = Polymer 1, # of Active Sites
MW = Polymer 1 Molecular Wt.
C = Concentration of Dye in Polymer 2
M = Partical Pressure of 0_2, mmHg
T = Blood Temperature
N = Compounded Noise Level

Control Factors, Modification of the L_{18} Array

Control factors and levels studied are shown in Table 1:

Table 1: Control Factors

	Factors	Level 1	Level 2	Level 3	Level 4	Level 5
A	Polymer 1 MW	.75	1.0	1.25	----	-----
B	Polymer 1 Reactive Sites	1.25	1.0	.75	----	-----
C	Dye/Polymer 1 Ratio	1.5	1.0	.5	----	-----
D	Sensor Thickness	Small	Med.	Large	----	-----
E	Dye Conc. in Polymer 2	1	2	3	----	-----
F	Overcoat Type	F1	F2	F3	F4	F5
G	Overcoat Size	Large	Norm.	----	----	-----

The composition of the oxygen sensor chemistry involves a two-step process; the first step entails attaching the fluorescing dye onto a special polymer (polymer 1). Factors studied were polymer l's molecular weight, the amount of chemical functionality present (polymer 1 reactive sites) and the molar ratio of dye to polymer (dye/polymer l ratio).

The second step combines and polymerizes polymer 1 (with fluorescing dye attached) with a second polymer while forming a thin disk. Factors studied were the thickness of the disk (sensor thickness) and the concentration of the polymer l/dye in the second polymer (dye concentration in polymer 2).

Also studied were the type of overcoat applied to the sensor, which optically isolates the sensor from outside light, as well as the size of the overcoat.

The L_{18} control factor array was modified to accept the levels of factors desired (compare Figure 3 to a standard L_{18} array). Columns 1 and 2 were combined to give a six-level column so that the five overcoat types could be studied. Level six of the combined column was run with a level 1 overcoat (Fl) and analyzed as a six-level factor as a self-check in the experiment (level 1 and level 6 should give near identical results – otherwise there is a problem). Column 5 was dummy-treated to reduce it from a three-level to a two-level column to accommodate the overcoat size factor.

Noise Factors, Compounding into "High" and "Low" Conditions

Noise factors studied were factors not easily controlled in our manufacturing process. They were compounded into "high" and "low" cases as seen in Table 2:

Table 2: Noise Factors Compounding

Factors	N1 "High"	N2 "Low"
Time at Elevated Temp.	0	5 Days
Exposure to Ambient Light	Minimal	2 Days
Sensor Thickness	+.001" from Target	–.001" from Target

Levels chosen for N1 were levels that would tend to increase the intensity of fluorescence of the sensor. N2 levels tend to decrease the sensor's intensity.

Experimental Methods – Experimental Procedure

The 18 variations of the oxygen chemistry indicated by the L_{18} array were formulated and sensors were produced from each variation. After subjecting some of each sensor type to both noise conditions, the sensors were ready to be tested in blood to determine their performance. Between three and eight sensors were tested at each $L_{18} x N_2$ design combination (we originally wanted to test eight at every combination, but experimental difficulties and manpower limitations prevented this).

Experimental Methods – Testing of Sensors at Various Signal Levels

The sensors were tested in what we call a "bovine blood loop" - an experimental apparatus that allows bovine blood to be circulated past numerous sensors at a time. This apparatus allows for precise control of the blood's oxygen concentration by use of blood oxygenators and standard gas mixtures, as well as temperature by using thermally controlled water baths. Bovine blood that has been adjusted chemically to simulate human blood is used routinely to test our products. Years of experience have shown good correlation between the bovine loop test and actual human clinical results.

The blood loops were run at the four concentrations shown below. At each concentration, the temperature of the blood was changed to within a degree of each of the three target temperatures (exact actual temperature was recorded). The intensity of fluorescence at each temperature point was measured by our monitors, which indicate a relative intensity from 0 to 1000 "counts." These intensity values were then used to calculate a signal-to noise ratio based on the theory presented in the next section.

Signal factors and their levels are shown in Table 3.

Table 3: Signal Factors

	Factors	Level 1	Level 2	Level 3	Level 4
M	Actual O_2 Concentration in Blood (mmHg)	42	76	160	228
M*	Temperature of Blood	18°C	28°C	37°C	---

Development of Signal-to-Noise Ratio

The signal-to-noise ratio that was developed to measure the goodness of each of the 18 design iterations was based on the ideal function of the sensor. In this section, I will start with an explanation of the ideal function of our sensor, and then show how we mathematically developed the signal-to-noise ratio used.

1. The ideal function of any measurement device is to indicate to the user a measurement that is exactly identical to the true value being measured (Figure 4) even under conditions of noise. Any deviation from this ideal function is an error in measurement.

Figure 4: General Ideal Function for O_2 Sensor

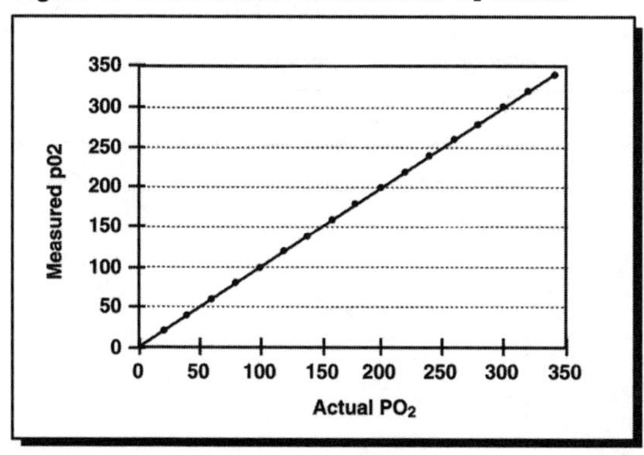

2. In the case of the fluorescing dye used as the active ingredient in our sensor, nature dictates that the inverse of the intensity of fluorescence emitted by the dye has a linear relationship with the concentration of oxygen present (oxygen quenches the dye's fluorescence, reducing intensity). It's this natural phenomenon that allows us to use this dye as the basis of our sensor. Figure 5 shows this relationship, and also that the measured concentration on the y-axis of Figure 4 has been replaced with the inverse of intensity.

Figure 5: O_2 Effect on Intensity of Fluorescent Dye

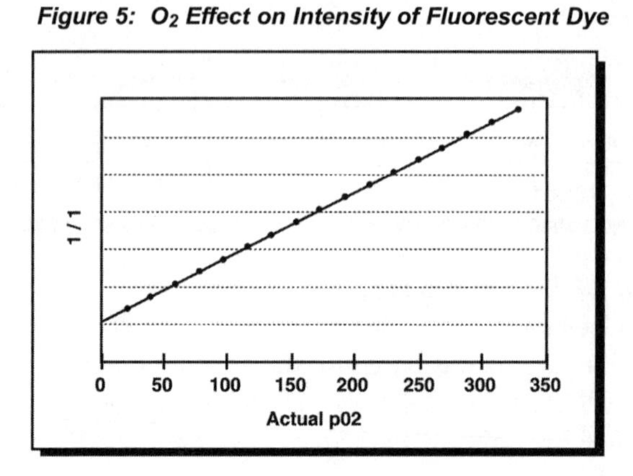

3. Given that our monitors measure only relative intensity, Figure 5 would only hold true for a single sensor on a single monitor. (Each monitor's excitation intensity varies slightly, and each sensor's thickness, etc., may also vary slightly, changing its intensity.) To allow us to compare many sensors' performance on the same scale we multiply by a correction factor specific to each sensor – the sensor intensity at zero oxygen, as shown in Figure 6.

Figure 6: Ratio Allows Sensor – Senor Comparison

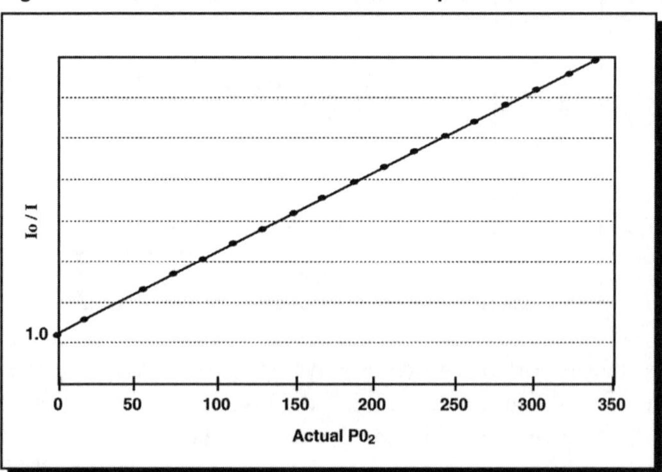

4. From Figure 6 we can write an equation to represent the fluorescence of the sensors when they are working ideally.

$$I_o/I = 1 + \beta(PO_2) \qquad (1)$$

Here, β represents the slope of the line depicted by Figure 6.

5. Now, this relationship holds true only at constant temperature, as the solubility of oxygen into the sensor disk has not been taken into the account. A way to compensate for temperature's effect is to calculate a correction for the temperature's effect on the sensor's slope.

$$\beta = \beta_{T_{ref}} + \beta^* (T - T_{ref}) \qquad (2)$$

where β^* is a constant, which we call the temperature coefficient of the sensor. T_{ref} is any reference temperature in the sensor's range of operation (37°C was chosen for this study).

6. Substituting the right side of Eq. (2) into Eq. (1) gives us the following relationship, which is graphically shown in Figure 7.

$$I_o/I = 1 + [\beta_{T_{ref}} + \beta^* (T - T_{ref})] (PO_2) \qquad (3)$$

Figure 7: Ideal function for O_2 Sensor Analysis

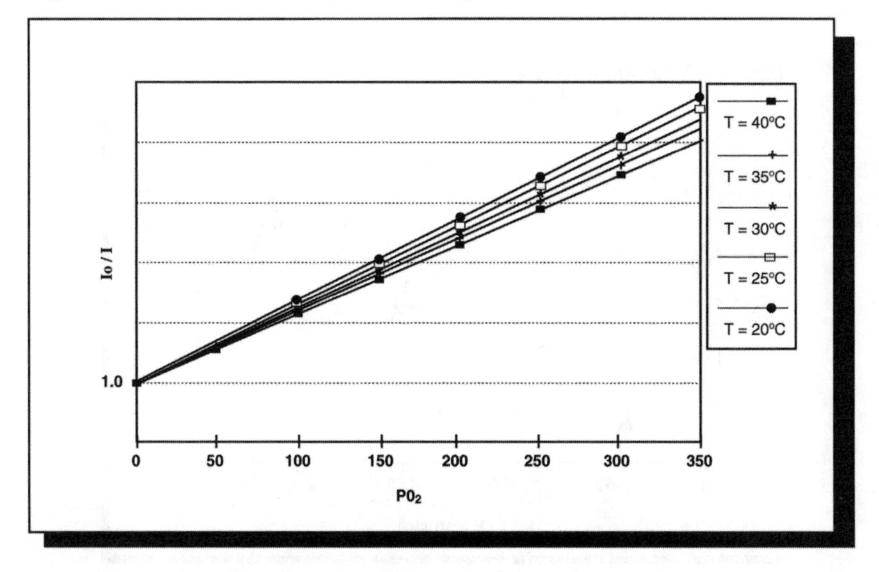

This equation describes the ideal function of our oxygen sensor. Any deviation from this ideal function constitutes an error in measurement.

7. The equation for I^o/I was rewritten as follows:

$$Y = [\beta + \beta^* (M^* - M_{ref}^*)] M \qquad (4)$$

Here, $Y = I_o / (I - 1)$.

Experimentally, the terms in the above equation represent the following:

Y	The measured value (calculated from the measured intensity and I^o).
β	The slope of the relationship between Y and the primary signal factor; partial pressure of oxygen, PO_2.
β^*	The slope of the relationship between β and temperature; the temperature effect on β.
M^*	Actual temperature, °C.
M_{ref}^*	Reference temperature, 37.0° C.
M	Value of primary signal factor, PO_2.

8. In this experiment, 12 combinations of T and M (see Table 3 and Figure 3) were run for each of the 18 design iterations. At each of the 12 combinations an intensity, and thus a Y, was measured and recorded for each sensor. These data are shown graphically in the appendices; iteration 5 is shown below. Notice iteration 5 generally follows the ideal relationship shown in Figure 7, but there is some deviation or error from the ideal.

Double Signal Analysis (Run 5)

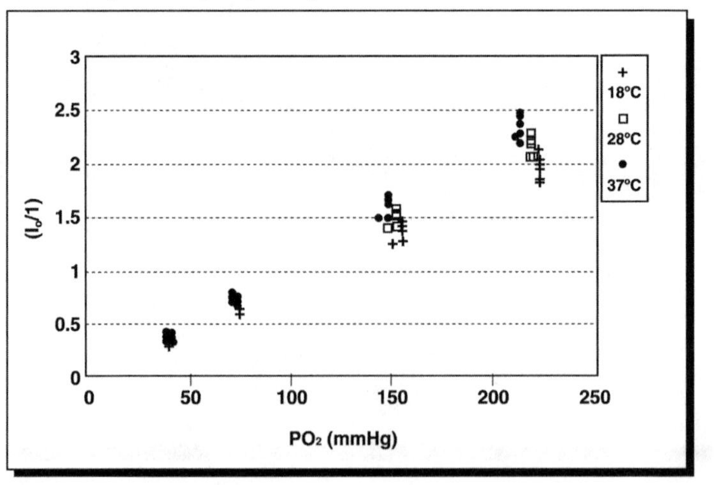

What was needed to calculate the integrity of each iteration is a signal-to-noise measurement that measures the deviation from the ideal in each of the experimental runs. This was accomplished as follows:

9. The equation for Y was modified by dividing each side of the equation by M, and then substituting Q for $(M^* - M^*_{ref})$:

$$Y/M = \beta + \beta^* Q \qquad (5)$$

Substituting X for Y/M results in a simple linear equation, as shown below, in which X = Y/M is the "response," β (slope of sensor) is the Y intercept, and $Q = T - T_{ref}$ is the signal.

$$\mathbf{X = \beta + \beta^{*}Q} \qquad \mathbf{(6)}$$

This equation is the basis for calculating signal-to-noise ratios, because as it uses a "reference point" relationship for temperature, and a "zero-point" relationship for the primary signal O_2 concentration.

10. A perfect sensor design would follow the above equation exactly, without any deviation or error from this relationship. The deviation a design exhibits can be calculated by dividing the total variation a sensor design exhibits, S_T, into two components: S_{β^*}, the "signal" portion of the variation, and S_e, the error portion of the variation. S_T, S_{β^*}, and S_e are calculated as follows:

$$S_T = \Sigma X_i - \frac{(\Sigma X_i)}{n} \qquad (7)$$

$$S_{\beta^*} = (1 / r) \{\Sigma [X_i - (Q_i - \overline{Q})]\}^2 \qquad (8)$$

$$S_e = S_T - S_{\beta^*} \qquad (9)$$

$$\text{where} \quad r = \Sigma (Q_i - \overline{Q})^2 \qquad (10)$$

11. The error variance, or "noise" component, of the signal-to-noise ratio is calculated as follows:

$$V_e = S_e / (n - 2) \qquad (11)$$

12. Finally, the signal-to-noise ratio is calculated:

$$S/N = 10 \log \frac{(1 / r) \; [S_{\beta^*} - V_e]}{V_e} \qquad (12)$$

13. The sensor's slope, β, and temperature coefficient, β*, can also be calculated as follows:

$$\beta^* = (1 / r) \; [S \, X_i \, (Q_i - \overline{Q})] \qquad (13)$$

$$\beta = \overline{X} - \beta^* \overline{Q} \qquad (14)$$

In the next section, a sample calculation is presented to elucidate the above theory.

Example Calculation – Run 5

a. Table 4 shows the raw data. There are seven sensors (5.1 to 5.7); the first four were exposed to noise level N1, the last three to noise level N2. Each sensor was tested at 12 combinations of M and M* (four M levels x three M* levels). After the 12th test point, the sensors were returned to the 1st test point as an experimental check, resulting in the 13 test points shown.

Table 4: Raw Data for Run 5

n	ID	Noise	M	M*	Y	n	ID	Noise	M	M*	Y
1	5.1	N1	41.24	18.6	0.373	53	5.5	N2	41.00	18.1	0.365
2	5.1	N1	223.56	16.8	2.031	54	5.5	N2	223.53	17.2	2.001
3	5.1	N1	74.40	18.6	0.704	55	5.5	n2	74.34	18.1	0.683
4	5.1	N1	155.96	18.1	1.436	56	5.5	N2	156.14	17.2	1.390
5	5.1	N1	40.59	27.7	0.403	57	5.5	N2	40.30	27.9	0.401
6	5.1	N1	219.45	27.8	2.279	58	5.5	N2	219.44	28.0	2.186
7	5.1	N1	73.17	28.0	0.750	59	5.5	N2	73.08	27.9	0.741
8	5.1	N1	153.31	27.9	1.564	60	5.5	N2	153.31	27.9	1.522
9	5.1	N1	39.45	37.5	0.442	61	5.5	N2	39.23	37.2	0.430
10	5.1	N1	213.67	37.1	2.456	62	5.5	N2	213.67	37.2	2.354
11	5.1	N1	71.12	37.7	0.814	63	5.5	N2	71.14	37.2	0.795
12	5.1	N1	149.25	37.2	1.689	64	5.5	N2	149.36	37.0	1.630
13	5.1	N1	41.28	17.9	0.346	65	5.5	N2	40.97	18.6	0.350
14	5.2	N1	41.14	18.6	0.311	66	5.6	N2	40.99	18.4	0.326
15	5.2	N1	223.39	17.4	1.831	67	5.6	N2	223.68	18.0	1.839
16	5.2	N1	74.41	18.5	0.589	68	5.6	N2	74.56	18.3	0.638
17	5.2	N1	156.08	17.5	1.268	69	5.6	N2	151.09	17.9	1.247
18	5.2	N1	40.46	28.0	0.334	70	5.6	N2	40.32	27.8	0.353
19	5.2	N1	218.95	28.8	2.060	71	5.6	N2	220.20	27.2	2.067
20	5.2	N1	73.16	28.1	0.667	72	5.6	N2	73.30	28.0	0.682
21	5.2	N1	153.07	28.6	1.397	73	5.6	N2	148.75	27.1	1.387
22	5.2	N1	39.40	37.2	0.361	74	5.6	N2	39.21	37.4	0.386
23	5.2	N1	214.12	36.5	2.193	75	5.6	N2	214.03	37.2	2.281
24	5.2	N1	71.30	37.0	0.697	76	5.6	N2	71.26	37.6	0.729
25	5.2	N1	149.68	36.4	1.494	77	5.6	N2	144.45	37.4	1.489
26	5.2	N1	41.12	19.0	0.304	78	5.6	N2	40.99	18.4	0.312
27	5.3	N1	41.14	18.6	0.366	79	5.7	N2	41.16	18.2	0.357
28	5.3	N1	223.39	17.4	1.978	80	5.7	N2	223.56	17.1	1.949
29	5.3	N1	74.41	18.5	0.664	81	5.7	N2	74.32	18.3	0.664
30	5.3	N1	156.08	17.5	1.385	82	5.7	N2	156.12	17.3	1.353
31	5.3	N1	40.46	28.0	0.400	83	5.7	N2	40.46	28.0	0.394
32	5.3	N1	218.95	28.8	2.199	84	5.7	N2	219.13	28.6	2.169
33	5.3	N1	73.16	28.1	0.738	85	5.7	N2	73.05	28.1	0.719
34	5.3	N1	153.07	28.6	1.546	86	5.7	N2	153.11	28.5	1.508
35	5.3	N1	39.40	37.2	0.433	87	5.7	N2	39.35	37.5	0.434
36	5.3	N1	214.12	36.5	2.343	88	5.7	N2	212.95	38.1	2.253
37	5.3	N1	71.30	37.0	0.791	89	5.7	N2	71.04	37.6	0.786
38	5.3	N1	149.68	36.4	1.611	90	5.7	N2	148.86	37.9	1.649
39	5.3	N1	41.12	19.0	0.366	91	5.7	N2	41.14	18.6	0.348
40	5.4	N1	40.87	19.2	0.386						
41	5.4	N1	223.11	19.1	2.140						
42	5.4	N1	74.34	19.3	0.703						
43	5.4	N1	155.73	19.2	1.463						
44	5.4	N1	40.26	27.6	0.401						
45	5.4	N1	219.58	28.0	2.273						
46	5.4	N1	73.23	27.7	0.747						
47	5.4	N1	153.28	28.0	1.572						
48	5.4	N1	39.24	36.7	0.417						
49	5.4	N1	213.65	37.4	2.411						
50	5.4	N1	71.10	37.8	0.804						
51	5.4	N1	149.14	37.4	1.678						
52	5.4	N1	40.88	19.0	0.369						

b. Table 5 shows the calculated data: X was calculated for each data point by dividing Y by M. For convenience, the numbers are multiplied by 1000 to make the numbers easier to work with. Also, Q was calculated by subtracting the reference temperature (37.0° C) from each measured temperature (M*). Further arithmetic manipulation gave X^2, $(Q - \bar{Q})^2$, and $X - (Q - \bar{Q})$. The data in each column were averaged or summed as necessary.

Table 5: Calculated Data for Run 5

n	Y	X	X^2	Q	$(Q-\bar{Q})^2$	$X-(Q-\bar{Q})$	n	Y	X	X^2	Q	$(Q-\bar{Q})^2$	$X-(Q-\bar{Q})$
1	0.373	9.05	81.97	−18.4	71.96	−76.80	53	0.365	8.89	79.07	−18.9	80.68	−79.87
2	2.031	9.09	82.57	−20.2	105.73	−83.43	64	2.001	8.95	80.15	−19.8	97.66	−88.47
3	0.704	9.46	89.46	−18.4	71.95	−80.23	55	0.683	9.19	84.37	−18.9	80.68	−82.51
4	1.436	9.21	84.79	−18.9	80.68	−82.71	56	1.390	8.90	79.21	−19.8	97.66	−87.95
5	0.403	9.92	98.45	−9.3	0.38	6.13	57	0.401	9.94	96.77	−9.1	0.67	8.13
6	2.279	10.38	107.82	−9.2	0.51	7.45	58	2.186	9.96	99.26	−9.0	0.84	9.14
7	0.750	10.25	106.12	−9.0	0.84	9.41	59	0.741	10.14	102.84	−9.1	0.67	8.29
8	1.564	10.20	104.02	−9.1	0.67	8.34	60	1.522	9.93	98.56	−9.1	0.67	8.12
9	0.442	11.20	125.41	0.5	106.53	116.66	61	0.430	10.96	120.04	0.2	102.37	110.85
10	2.456	11.49	132.08	0.1	100.35	115.13	62	2.354	11.02	121.37	0.2	102.37	111.46
11	0.814	11.44	130.98	0.7	112.73	121.62	63	0.796	11.18	124.89	0.2	102.37	113.07
12	1.689	11.32	128.09	0.2	102.37	114.51	64	1.630	10.91	119.04	0.0	96.36	108.21
13	0.346	8.38	70.22	−19.1	84.32	−76.95	65	0.350	8.55	73.02	−18.4	71.95	−72.48
14	0.311	7.56	57.21	−18.4	71.95	−64.16	66	0.326	7.96	63.37	−18.6	75.38	−69.12
15	1.831	8.20	67.19	−19.6	93.75	−79.37	67	1.839	8.22	67.58	−19.0	82.49	−74.66
16	0.589	7.92	62.66	−18.5	73.66	−67.94	68	0.638	8.55	73.13	−18.7	77.13	−75.10
17	1.268	8.12	65.96	−19.5	91.82	−77.83	69	1.247	8.25	68.07	−19.1	84.32	−75.76
18	0.334	8.26	68.26	−9.0	0.84	7.58	70	0.353	8.77	76.86	−9.2	0.51	6.29
19	2.060	9.41	88.52	−8.2	2.95	16.16	71	2.057	9.39	88.14	−9.8	0.01	1.10
20	0.662	9.05	81.93	−8.9	1.04	9.21	72	0.682	9.31	86.59	−9.0	0.84	8.54
21	1.397	9.13	83.33	−8.4	2.30	13.85	73	1.387	9.33	86.98	−9.9	0.00	0.16
22	0.361	9.17	84.14	0.2	102.37	92.80	74	0.386	9.84	96.73	0.4	106.45	101.47
23	2.193	10.24	104.91	−0.5	88.69	96.46	75	2.281	10.66	113.54	0.2	102.37	107.81
24	0.697	9.78	95.57	0.0	98.36	96.96	76	0.729	10.23	104.65	0.6	110.62	107.59
25	1.494	9.98	99.65	−0.6	86.82	93.01	77	1.489	10.31	106.30	0.4	106.45	106.38
26	0.304	7.39	54.62	−18.0	65.33	−59.74	78	0.312	7.62	58.03	−18.6	75.38	−66.14
27	0.366	8.91	79.36	−18.4	71.96	−75.56	79	0.367	8.68	75.31	−18.8	78.90	−77.06
28	1.978	8.86	78.43	−19.6	93.75	−85.75	80	1.949	8.72	76.00	−19.9	99.65	−87.03
29	0.664	8.92	79.59	−18.5	73.66	−76.56	81	0.664	8.94	79.88	−18.7	77.13	−78.49
30	1.386	8.88	78.78	−19.5	91.82	−86.06	82	1.363	8..86	75.06	−19.7	96.70	−84.75
31	0.400	9.89	97.84	−9.0	0.84	9.06	83	0.394	9.75	95.06	−9.0	0.84	8.95
32	2.199	10.04	100.89	−8.2	2.95	17.25	84	2.199	9.90	97.96	−8.4	2.30	15.02
33	0.738	10.06	101.65	−8.9	1.04	10.26	85	0.719	9.86	96.94	−8.9	1.04	10.02
34	1.546	10.10	101.99	−8.4	2.30	15.33	86	1.508	9.86	97.06	−8.5	2.01	13.97
35	0.433	10.96	120.67	0.2	102.37	111.14	87	0.434	11.02	121.55	0.5	106.53	114.85
36	2.343	10.94	119.75	−0.5	88.69	103.06	88	2.253	10.58	111.94	1.1	121.39	116.57
37	0.791	11.09	123.06	0.0	96.36	110.02	89	0.786	11.06	122.29	0.6	110.62	116.31
38	1.611	10.76	115.78	−.06	86.82	100.26	90	1.649	11.06	122.68	0.9	117.02	119.81
39	0.366	8.89	79.07	−18.0	65.33	−71.87	91	0.348	8.47	71.73	−18.4	71.96	−71.84
40	0.386	9.44	89.12	−17.8	62.13	−74.41							
41	2.140	9.59	92.00	−17.9	63.72	−76.56	Avgs:		9.67		-0.92		
42	0.703	9.46	89.42	−17.7	60.57	−73.59	Sums:		879.57	8696.06		5683.21	623.25
43	1.463	9.40	88.28	−17.8	62.13	−74.06							
44	0.401	9.96	99.03	−9.4	0.27	5.15							
45	2.273	10.35	107.12	−9.0	0.84	9.50							
46	0.747	10.20	104.07	−9.3	0.38	6.30							
47	1.572	10.26	105.21	−9.0	0.84	9.41							
48	0.417	10.62	112.83	−0.3	92.50	102.16							
49	2.411	11.29	127.40	0.4	106.45	116.45							
50	0.804	11.31	127.96	0.8	114.87	121.25							
51	1.678	11.25	126.52	0.4	106.45	116.06							
52	0.369	9.02	81.30	−18.0	65.33	−72.87							

c. Using Eqs. (7) to (10) and the column averages and sums from Table 5, S_T, $S_{\beta*}$, and S_e were calculated:

$r = 5683.21$ (column sum from Table 5)

$S_T = 94.50$

$S_{\beta*} = 68.35$

$S_e = 94.50 - 68.35$
$S_e = 26.15$

d. The error variance was calculated per Eq. (11):

$V_e = 26.15 / 89 = .294$

e. S/N ratio was calculated per Eq. (12):

f. $\beta*$ and β were calculated per Eqs. (13) and (14) (values were divided by 1000 since original X values were multiplied by 1000 for convenience):

$\beta* = [(1 / 5683.21)(623.25)] / 1000$
$\beta* = 1.10 \times 10^{-4}$

$\beta = [9.67 - (-9.92(1.10 \times 10^{-4} \times 1000)] / 1000$
$\beta = .0108$

g. Similar calculations were performed for each of the other 17 iterations, resulting in the data given in Table 6.

Results and Discussion

Table 6 shows the S/N ratios calculated for each of the 18 design iterations as well as the sensor's average slope (β) and temperature coefficient ($\beta*$). Runs 2, 8 and 18 had the best S/N ratios, while runs 3, 6 and 16 had the worst. The S/N ratios represent the raw data very well as graphs of the raw data (see appendices), and show that runs 2, 8 and 18 fit the ideal function very well, whereas runs 3, 6 and 16 deviate significantly.

Table 6: Experimental Results

L_{18} Run	S/N Ratio	β	β*
1	−18.14	.0108	1.18×10^{-4}
2	−12.28	.0109	1.01×10^{-4}
3	−20.66	.0103	0.97×10^{-4}
4	−14.77	.0101	1.07×10^{-4}
5	−13.91	.0108	1.10×10^{-4}
6	−21.09	.0112	1.01×10^{-4}
7	−13.72	.0096	1.10×10^{-4}
8	−5.29	.0109	1.03×10^{-4}
9	−13.53	.0101	1.06×10^{-4}
10	−16.90	.0113	1.19×10^{-4}
11	−18.22	.0105	1.09×10^{-4}
12	−13.42	.0114	1.24×10^{-4}
13	−17.72	.0109	0.97×10^{-4}
14	−14.03	.0106	1.08×10^{-4}
15	−15.50	.0104	1.10×10^{-4}
16	−19.89	.0100	0.97×10^{-4}
17	−16.37	.0108	0.88×10^{-4}
18	−12.16	.0118	1.06×10^{-4}

An analysis of variance was performed for the S/N, slope, and temperature coefficient data to determine which factors influenced each significantly. From these analyses, optimum levels of each factor were chosen.

Analysis of Variance (S/N Data)

Table 5b shows the results of ANOVA (analysis of variance). The effects of the significant factors are shown in Figures 8 and 9.

Table 5b: ANOVA or S/N Ratio

Source	df	S	V	F	S'	rho%
Overcoat Type	5	94.68	18.94	2.77	60.53	22.9
Thickness	2	18.01	9.00	Pooled		Pooled
Dye/Polymer 1 Ratio	2	38.47	19.23	2.82	24.81	9.4
Overcoat Size	2	48.13	24.06	3.52	34.47	13.0
Polymer 1 Sites	2	13.08	6.54	Pooled		Pooled
Polymer 1 MW	2	9.87	4.93	Pooled		Pooled
Dye Conc. in Polymer 2	2	42.02	21.01	3.08	28.36	10.7
Pooled Error	4	40.96	6.83		116.1	43.9
TOTAL	17	277.30	16.31			100.0

Figure 8: Effects of Overcoat Type and Dye Concentration on S/N Ratio

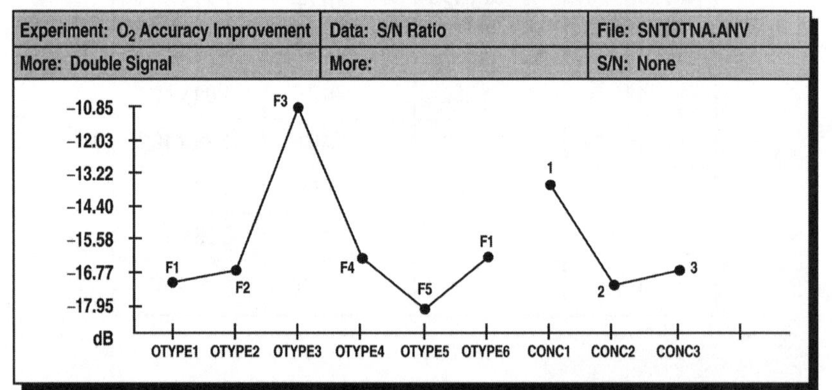

Figure 9: Effects of Dye/Polymer 1 Ratio and Overcoat Size on S/N Ratio

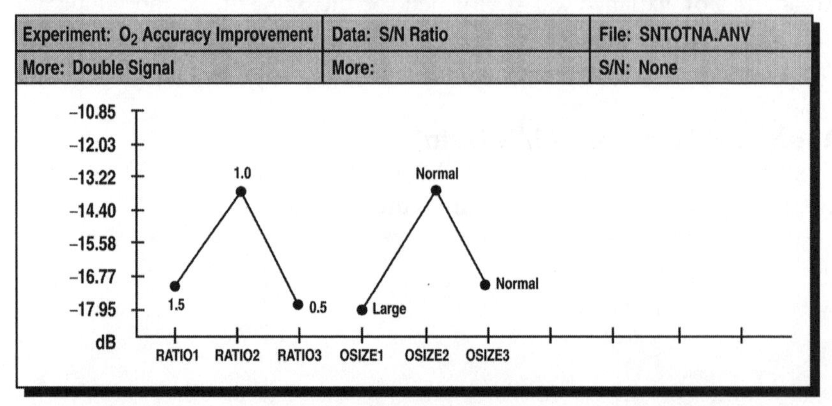

The type of overcoat used to fabricate sensors had the largest effect (22.9% contribution) on the S/N ratio, and thus the accuracy of the sensor. Figure 8 shows us that type F3 was significantly better than other types (6 dB better), and also that the other four types were approximately equivalent. Also, Figure 8 shows that overcoat type levels 1 and 6 (both type Fl) had nearly identical level averages, which indicates the experimental design worked well and nothing unexpected happened to spoil the experiment or analysis such as a strong interaction, data entry errors or a testing error.

Dye concentration and the dye/polymer 1 ratio also had significant effects, with 1 and 1.0 respectively, being optimum levels.

The overcoat size effect is suspect because levels 2 and 3, which were (supposedly) identical, gave different results. I have not been able to determine why this is so. I most suspect some experimental errors in fabricating the overcoats as this was not a well-defined process at the time of the experiment. Perhaps also some interactions between control factors may exist (confirmation runs showed the interactions were not serious).

The effect of thickness, active sites, and molecular weight had no effect on the S/N ratio, and thus the accuracy of the sensor. This is an important finding, as this gives us the latitude to manufacture within wider variances without affecting the performance of our product.

Analysis of Variance – Slope (Sensitivity)

Table 6b shows that a number of factors significantly affect the slope of the sensor (β) from a statistical standpoint (F values large), particularly the amount of reactive sites on polymer 1. However, from a practical standpoint the level average changes seen in this experiment (.010 – .011) are small (see Figures 10 to 11). Therefore I've concluded that the factors studied, within the ranges studied, don't significantly affect the slope of the sensors.

This is another important result, as it gives us the freedom to pick the levels of the factors studied with respect to attributes other than accuracy or cost reduction, and also gives guidelines about how tightly these factors must be controlled during manufacturing.

Table 6b: ANOVA or Slope

Source	df	S	V	F	S'	rho%
Overcoat Type	5	1.25E-6	2.50E-7	9.19	1.12E-6	21.5
Thickness	2	5.54E-7	2.77E-7	10.18	5.00E-7	10.8
Dye/Polymer1 Ratio	2	6.14E-7	3.07E-7	11.29	5.60E-7	10.8
Overcoat Size	2	5.44E-8	2.72E-8			Pooled
Polymer 1 Sites	2	1.72E-6	8.62E-7	31.67	1.67E-6	32.3
Polymer 1 MW	2	5.44E-8	2.72E-8			Pooled
Dye Conc.	2	9.24E-7	4.62E-7	16.99	8.70E-7	16.8
Pooled Error	4	1.09E-7	2.72E-8		4.63E-7	8.9
TOTAL	17	5.18E-6	3.05E-7			100.0

Figure 10: Analysis of Variance

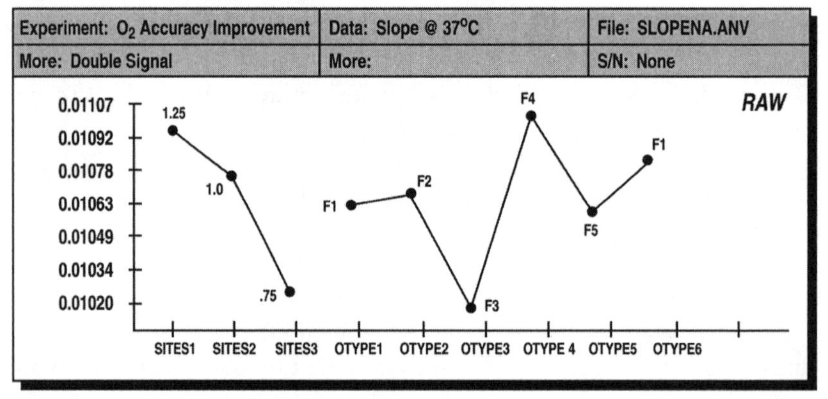

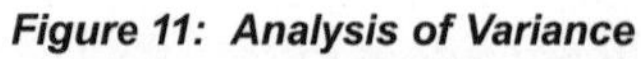
Figure 11: Analysis of Variance

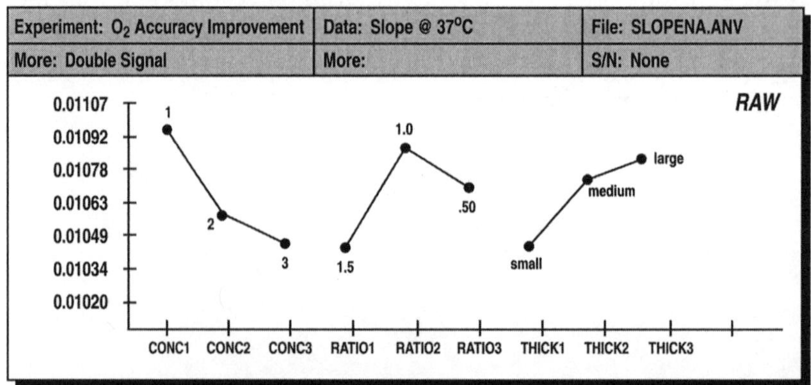

Analysis of Variance – Temperature Coefficient

Due to the sensitivity of this information, the analysis and associated findings were not included.

Selection of Optimum Levels

Given that the factors and ranges studied had little effect on slope, I chose optimum levels based entirely on the S/N analysis. Levels of factors found to be insignificant in the S/N analysis were chosen based on what works best for manufacturing the sensor.

The levels of the first three factors listed in Table 7 were chosen based on the S/N analysis above. Polymer l's active sites and molecular weight were chosen to reduce manufacturing costs. Sensor thickness was chosen at the midpoint to give us manufacturing flexibility. It was also decided not to change the overcoat size because it showed no conclusive benefit.

Table 7: Optimum Design

Factors	Optimum Levels
Overcoat Type	F3
Dye conc. in Polymer 2	1
Dye/Polymer 1 Ratio	1.0
Sensor Thickness	Medium
Polymer 1 Amount of Reactive Sites	1.0
Polymer 1 MW	1.0
Overcoat Parameter	Normal

Confirmation runs were carried out at the above conditions. Results are presented in the next section.

CONFIRMATION RUNS

Using ANOVA-TM software, predictions were made for the S/N ratio, slope and temperature coefficient at optimum design conditions. As can be seen in Table 8, the confirmation sensors performed within predicted ranges.

Of more importance is the 9-dB improvement in S/N ratio when compared to the controls, which are representative of our current sensor design.

The 9-dB improvement has resulted in a threefold improvement in the sensor's accuracy.

Table 8: Confirmation Run Data

	Predicted	Actual	Controls
S/N Ratio	–8.3 +/– 5.2	–12.7	–21.6
Slope (β)	.0111 +/– .0004	.0110	.0071
Temp. Coeff. (β*)	1.14×10^{-4} +/– $.06 \times 10^{-4}$	1.18×10^{-4}	0.88×10^{-4}

CONCLUSION

1. The type of overcoat, dye concentration in polymer 2 and the dye/polymer 1 ratio affect the accuracy of the oxygen sensor studied. Setting these factors to their optimum levels improved the accuracy of the sensor threefold (sensor error is one-third of original).

2. Polymer l's number of active sites and molecular weight, as well as sensor thickness did *NOT* affect accuracy, and therefore can be adjusted to improve the manufacturing of the sensors.

This case study is contributed by John T. Bires, CDI/3M Health Care, California. A special thank you is extended to Christa Ruiz, Chemical Technician at CDI/3M for her fine work in producing the experimental batches and sensors for this study, and for her diligence in running the many bovine blood loops, each of which is quite an involved and exhausting procedure.

CHAPTER FIVE

Optimization of Nickel-Cadmium Battery Operation Management Using Robust Design

JET PROPULSION LABORATORY, NASA AND CALIFORNIA INSTITUTE OF TECHNOLOGY – USA

EXECUTIVE SUMMARY

BACKGROUND

In recent years, following several spacecraft battery anomalies, it was determined that managing the operational factors of the NASA flight NiCd rechargeable battery was very important in order to maintain spaceflight battery nominal performance. Hence, NASA had to call upon JPL to initiate studies and analysis in order to establish an operations management protocol for these batteries. The optimization of existing flight battery operational performance was viewed as something new for Taguchi Methods™ of application.

PROCESS

The evaluation, qualification and operations management of secondary batteries for NASA space vehicles are involved and constitute a very lengthy process. Rechargeable battery evaluation requires tens or even hundreds of cycles. Testing for the performance effects of these parameters could be a never-ending task. In this particular case, a full factorial with five factors at four levels would have required 1024 experiments. Since each experiment must be performed at the given levels for 60 cycles (approximately four days), a total of 4096 days or 11.2 years of experimentation would have been required had this approach been adopted. With Taguchi Methods, the best results were obtained with only 16 experiments. Because of time constraints, the dynamic S/N ratio approach could not be used. In this experiment, the S/N ratio in the case of "nominal the best" approach was used.

BENEFITS

- The application of Taguchi Methods resulted in cost savings of over 400% and over 300% reduction in experimental time and at the same time improved battery voltage performance over 96% (as compared to existing time).

- Substantial amount of time was saved by resorting to this approach.
- This innovative application of Taguchi's robust design is viewed as a new technology of applying this modern engineering design optimization technique to the operational optimization of existing spaceflight battery in order to improve battery life nominal performance and thus extend spacecraft life.

CASE STUDY

INTRODUCTION

Nickel cadmium rechargeable batteries are currently used for an entire class of NASA observatory spacecraft including GRO, UARS, EUVE and TOPEX/Poseidon. Optimum levels of onboard spacecraft battery operation performance were determined to extend the life of these batteries and thus the life of NASA spacecraft. In recent years, several spacecraft NiCd battery anomalies occurred that drastically affected spacecraft life. This prompted NASA to call upon JPL to initiate studies and analysis in order to establish an operations management protocol for these batteries.

The evaluation, qualification and operations management of secondary batteries for NASA space vehicles are involved and constitute a very lengthy process. There are many variables and levels of each variable which affect the overall reliability and performance of batteries. Rechargeable battery evaluation requires tens or even hundreds of cycles. Testing for the performance effects of these parameters could be a never-ending task.

OBJECTIVES

NASA was concerned about the performance of the existing onboard batteries. The challenge faced was to design a protocol for the battery operation process to achieve life performance optimization in the shortest time possible with minimum cost while significantly improving battery performance. At first, since this was not viewed as a classical product or manufacturing process design optimization, no relation was seen to Dr. Genichi Taguchi's method of robust design. Nevertheless, at a closer look, it became obvious that the optimization of an operations process, in this case a rechargeable battery operation, is no different than optimization of any process.

A team of battery experts was formed at JPL to perform a study of battery operational optimization using the old methods. Based on practical experience, it was determined that controlling the recharge fraction of flight batteries in operation was important to maintain nominal performance. The recharge fraction is one of the parameters used to determine battery overcharge. The recharge fraction is normally derived on an orbit basis and there are several operating factors that influence it. The factors influencing the recharge fraction are:

1. Charge current during peak power tracking (peak charge current)
2. Battery depth of discharge
3. Operating temperature
4. Orbit duration
5. V/T level of charging

The best experiments were performed where the above five factors were set at estimated levels. After over a year, the best battery operational performance was established. Figure 1 describes the cell voltage divergence profile optimization using the old method.

After analyzing the battery performance of Figure 1, it was soon realized that the best battery operations management was far from an ideal functional performance as shown in Figure 2. A more quantifiable experimentation and analysis for further battery operational optimization was needed.

Parameter Design Approach

In performing battery operations management optimization in the past, JPL used the classical approach to experimentation, which is to modify one parameter and keep the rest of the parameters fixed. Most often, his old method requires considerable time and resources in order to attain an acceptable performance.

Figure 1: Voltage Profile Prior to Applying Robust Design

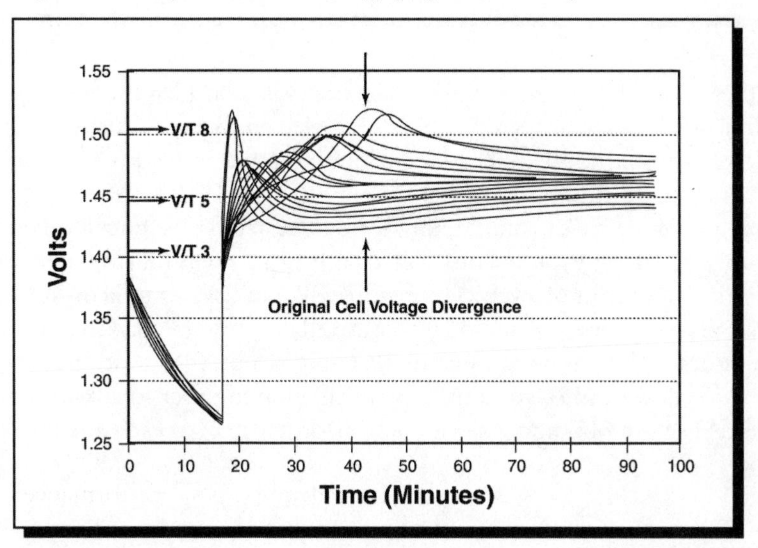

Figure 2: Ideal Voltage Profile

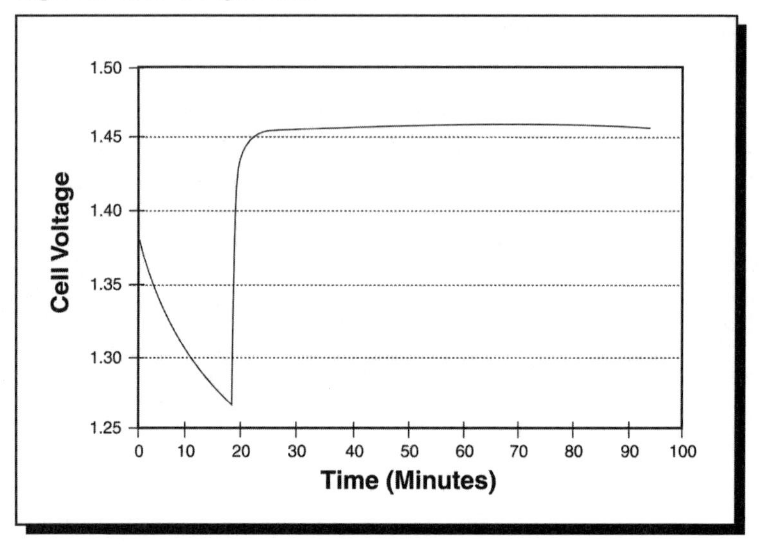

Factors and Levels

Factors	Levels			
	1	2	3	4
Peak Charge Current (A)	10	20	30	40
DOD (%)	5	10	15	25
Temperature (°C)	0	5	10	15
Orbit Duration (min)	90	100	110	120
V/T Level	2	3	4	5

For these reasons, designed experimentation was considered next. Each of the five previously considered factors was selected to perform at four different levels as listed to the right in the table above and in Figure 3.

In this particular case, a full factorial with five factors at four levels would have required 1024 experiments. Since each experiment must be performed at the given levels for 60 cycles (approximately four days), a total of 4096 days or 11.2 years of experimentation would have been required had this approach been adopted. It was very obvious that it was not very cost effective for NASA to allow over 11 years of experimentation in order to obtain the data and establish the optimum operational performance of these batteries.

Novel battery management techniques had to be implemented to quickly recover spaceflight battery performance. For this reason a NASA battery testbed was established to systematically evaluate various battery management techniques.

This was the time when Taguchi Methods of robust design were first considered in order to improve battery life by optimizing the battery operations process. To quickly determine which of the above factors needed to be operated at what levels and to influence the battery recharge fraction the most, fractional factorial techniques were considered.

The proposed test articles were three existing 22-cell nickel-cadmium batteries available at JPL. Two batteries were approximately nine years old and had been used on the GRO and TOPEX/Poseidon missions as "test and integration" batteries. The third battery was assembled with cells from four different manufacturing lots after the cells were cycled for several hundred cycles. Thus, there was plenty of product-to-product noise.

Experimental Layout, Results and Analysis

In setting up the Taguchi designed experiment, the five above-described factors each at four levels were studied. A modified L_{16} orthogonal array was selected for this experiment which allowed evaluation of the five factors at four levels each. With each experiment performed 60 times, the total duration of this experiment was reduced from the initial 11.2 years to only 10 weeks.

Even though a significant signal factor was identified, due to time and cost constraints a static robust design was performed. The macromodeling or P-diagram approach is described in Figure 3.

Figure 3: Ni-Cd Battery P-Diagram

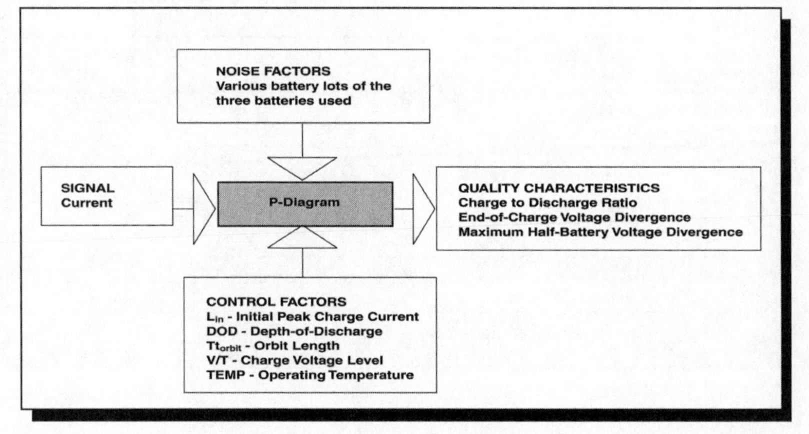

Figure 4: Experiment Setup and Output Measurement

Variable Settings						Averages																	
						Charge to Discharge Ampere-Hour Ratio			End-of-Charge Current (Amperes)			End-of-Charge Range Cell Voltage (MV)			Maximum Range of Cell Voltages (MV)			End-of-Charge Half-Battery Delta (MV)			Maximum Half-Battery Voltage Delta (MV)		
Experiment	Temperature (°C)	Orbit Duration (minutes)	DOD Based on 130AH Nameplate	Inrush Current	Max. Volts per Cell	A	B	C	A	B	C	A	B	C	A	B	C	A	B	C	A	B	C
11	0	100	15%	90	1.46	1.01	1.02	1.02	0.8	0.7	13.8	45.6	15.6	28.1	50.9	50.9	24.2	–4.3	4.6	–1.1	19.8	39.7	24.9
16	0	110	25%	120	1.42	1.00	1.00	1.01	2.7	2.7	1.8	2.7	5.1	17.4	21.9	28.5	18.0	–2.9	0.1	–3.8	13.0	10.5	11.1
6	0	120	10%	60	1.44	1.02	1.02	1.03	0.6	0.6	0.4	13.3	29.3	12.9	34.8	50.0	32.1	–25.9	23.4	3.0	90.9	29.0	37.1
1	0	90	5%	30	1.4	1.01	1.01	1.01	0.4	0.4	0.3	12.4	21.4	13.5	37.6	52.5	20.3	–25.0	15.5	0.2	73.7	26.1	32.9
5	5	110	5%	60	1.45	1.08	1.08	1.16	0.4	0.4	0.4	46.7	74.1	24.3	132.9	136.5	41.0	–96.1	50.5	–8.0	327.6	88.6	65.7
12	5	90	25%	90	1.43	1.00	1.00	1.01	6.0	6.0	3.8	4.3	9.8	23.1	30.9	27.3	23.4	31.4	31.4	31.4	21.4	15.9	12.8
15	5	120	15%	120	1.39	1.01	1.01	1.01	1.1	1.1	0.6	2.3	6.1	7.6	36.3	39.6	24.5	–2.4	4.7	3.3	15.2	13.0	19.9
2	5	100	10%	30	1.41	1.01	1.01	1.01	1.2	1.2	0.6	5.2	11.3	11.5	27.0	38.2	28.5	–9.0	9.6	3.3	15.5	11.2	38.2
8	10	100	25%	60	1.38	0.98	0.98	0.99	5.2	5.2	5.3	3.4	3.9	5.5	6.6	7.5	29.5	–5.6	–1.1	6.2	4.2	3.2	20.1
9	10	120	5%	90	1.4	1.06	1.05	1.05	0.4	0.4	0.2	4.7	5.1	14.3	38.7	40.9	33.6	11.7	1.0	–1.6	23.8	16.4	54.5
14	10	90	10%	120	1.44	1.05	1.05	1.11	1.2	1.2	0.9	29.3	50.6	22.8	56.1	67.5	34.9	–44.7	35.7	–7.0	111.7	44.4	45.0
3	10	110	15%	30	1.42	1.01	1.01	10.3	1.8	1.8	0.9	14.3	21.3	17.4	19.9	26.5	22.4	–20.4	12.0	–0.8	39.5	12.0	35.4
4	15	120	25%	30	1.425	0.99	0.99	0.99	9.1	9.1	9.8	8.0	5.8	21.2	15.0	15.7	23.9	–1.5	–2.2	1.8	4.1	4.7	18.2
7	15	90	15%	60	1.385	1.01	1.01	1.01	1.9	1.9	1.7	3.7	3.3	5.6	21.9	28.7	20.7	2.2	1.4	3.2	4.2	4.2	20.9
10	15	110	10%	90	1.365	1.01	1.01	1.01	0.7	0.7	0.5	2.6	2.4	2.9	26.9	26.6	23.1	0.1	0.7	2.2	0.2	4.6	25.2
13	15	100	5%	120	1.405	1.07	1.07	1.10	0.8	0.8	0.5	2.7	3.8	16.6	39.0	38.6	36.0	4.7	–4.5	–.09	19.7	24.1	63.0
Ver. 1	5	120	25%	90	1.410	1.00	1.00	1.01	2.8	2.8	1.5	2.1	8.4	10.9	25.1	23.1	25.5	–4.3	4.8	0.3	-45.9	-13.9	12.7
Ver. 2	5	120	25%	90	1.410	1.00	1.00	1.01	3.0	3.0	1.6	3.0	5.2	11.6	29.2	25.3	21.5	–1.3	1.5	–2.1	-48.0	-14.0	22.9
Ver. 3	15	120	25%	90	1.410	1.01	1.02	1.03	2.6	2.6	1.5	4.5	8.2	14.0	26.4	23.8	21.5	–1.9	3.5	–1.3	-48.1	-8.7	34.4
UARS, Ref. 1	3	95	20%	102	1.460	1.02	1.02	1.04	2.4	2.3	1.5	21.0	36.6	10.4	39.0	57.3	41.0	–35.8	35.7	8.3	-72.0	47.8	48.4
UARS, Ref. 2	3	95	10%	102	1.460	1.03	1.04	1.09	1.0	1.0	0.8	33.3	40.7	14.9	89.3	98.4	31.7	–37.6	9.3	2.7	63.7	85.2	63.7
Ver. 4	15	120	25%	90	1.385	1.01	1.01	1.01	2.1	2.1	1.7	1.5	2.2	10.4	14.0	12.1	20.9	0.3	1.1	–0.6	-26.2	–9.9	18.0

Figure 5: Signal-to-Noise Ratio for Recharge Fraction

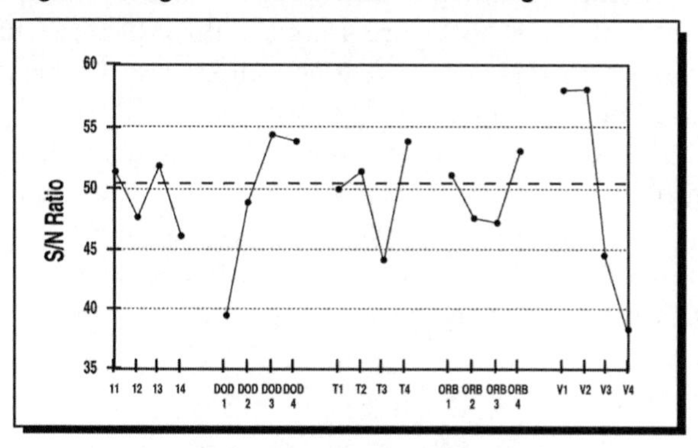

V/T & DOD have the largest influence on recharge fraction.
Peak charge current has the least influence.
High DOD and low V/T level result in worst recharge fraction.

Figure 6: Signal-to-Noise Ratio for End-of-Charge Divergence

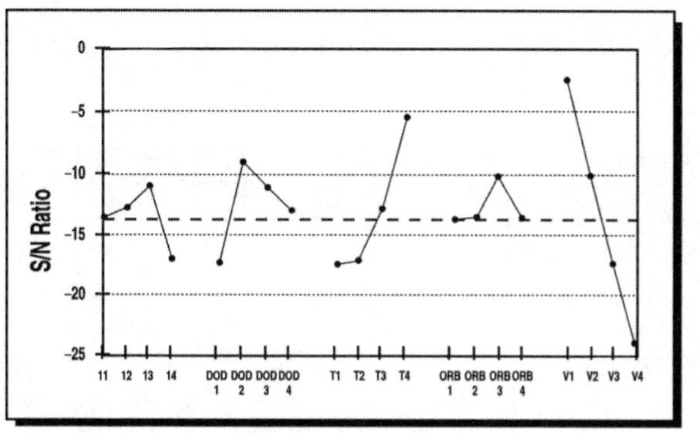

Figure 7: Signal-to-Noise Ratio for Maximum Half-Battery Divergence

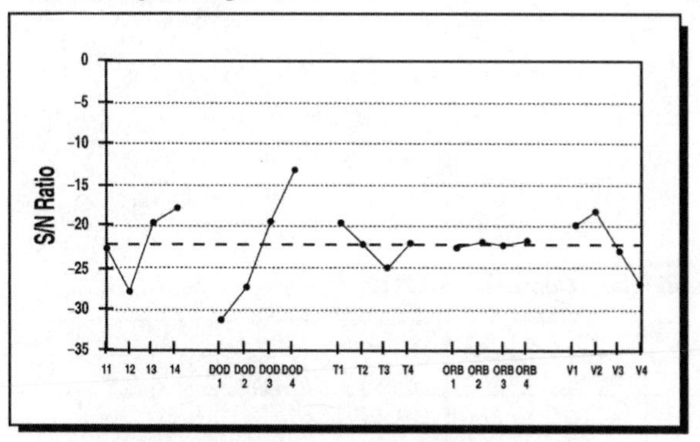

DOD is the worst influential parameter for half-battery divergence.
Orbit length has no influence on half-battery divergence.
High DOD results in the lowest half-battery divergence.

The setup of the experimentation and output measurements is described in Figure 4. The A, B and C in Figure 4 arc the three batteries under experimentation. It is worth mentioning that six output measurements or quality functions ranked in order of importance were recorded (see Figure 4). The ANOVA-TM professional software package was used to analyze the data. Signal-to-noise analysis was performed for "nominal the best" case. Sensitivity analysis was performed for all six measurements and response graphs are shown here for only three. Recharge fraction, end of charge divergence and; max-half-battery divergence are shown in Figures 5, 6, and 7.

CONFIRMATION

The operation optimization was performed against the first quality characteristic "recharge fraction" with the other five factors being used only to influence the factor level selection for the process's average prediction. Cost was not considered the factor levels.

The suggested parameter combination for the best performance was as follows:

Parameters	Verification Conditions
Peak Charge (Amp)	30
DOD (%)	25
Temperature (°C)	5
Orbit duration (Min)	120
V/T Level	3

The projected S/N ratio was 73.64 dB. The mean was T = 54.08 dB; thus there was a delta increase of 19.56 dB. Verification comparison data are shown in Figure 8.

Figure 8: Verification Comparison Data

Batt#	C/D			EOC Div.			Max. Half-Batt. Div.		
	A	B	C	A	B	C	A	B	C
Initial	1.03	1.04	1.09	–37.60	9.30	2.70	63.70	85.20	65.20
Ver. 1	1.00	1.00	1.01	–4.30	4.80	0.30	–45.90	–13.90	12.70
Ver. 2	1.00	1.00	1.01	1.30	1.50	–2.10	–48.00	–14.00	22.90

Figure 9 describes the voltage divergence profile after applying robust design. Comparing the profiles of the Figure 1 and Figure 9 graphic representations, before and after using robust design, the performance improvement was quite remarkable and was evaluated at over 96% improvement. This performance was superior to the projected improvement.

Figure 9: Voltage Profile After Applying Robust Design

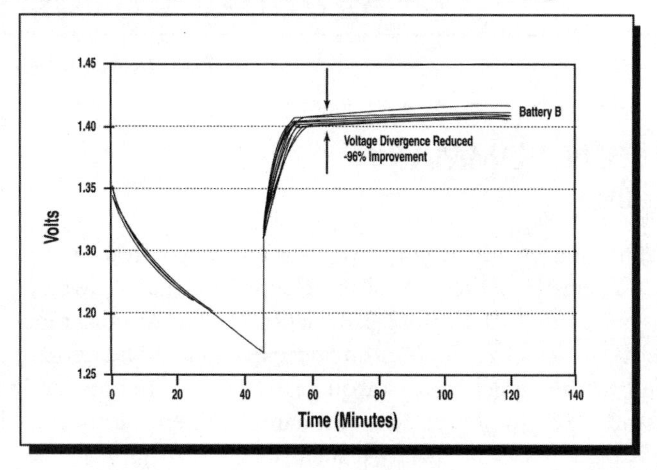

Conclusion

The excellent results of the application of the Taguchi Methods of robust design have assisted the power subsystem and battery analysis experts in determining the appropriate protocol for flight NiCd operations management for various current and future missions.

Results obtained using the old way of performing battery operations management were compared to the results obtained using the robust design approach.

The application of Taguchi Methods resulted in cost savings of over 400% and over 300% reduction in experimental time. At the same time it improved battery voltage performance over 96%.

This innovative application of Taguchi's robust design is viewed as a new technology in the application of the modern engineering design optimization technique to the operational optimization of existing spaceflight batteries in order to improve battery life nominal performance and thus extend spacecraft life.

This case study is contributed by Julian Blosiu, Frank Deligiannis, Salvador Di Stefano of Jet Propulsion Laboratory, NASA and California Institute of Technology, USA.

CHAPTER SIX 

Engine Idle Quality Robustness

FORD MOTOR COMPANY – USA

EXECUTIVE SUMMARY

BACKGROUND

The inherent random variation of engine combustion from one engine cycle to another results in indicated mean effective pressure (IMEP) variations. These variations in IMEP cause a deviation in the normal motion of the engine about its axis, resulting in engine vibrations. This vibration is then transmitted to the vehicle body through the engine mounts, and ultimately felt as disturbances by the vehicle occupants. When the magnitude of the vibration is high, poor idle quality results.

With increasing quality demands by the customers, the idle quality issue is of major concern. It was necessary to find a solution that will improve idle quality by minimizing the effect of the noise factors in the field.

PROCESS

Instead of following the common one-factor-at-a time method, technically more efficient robust design methodology was used. Eight control factors were considered under customer usage conditions (engine mileage). Eighteen design configurations were tested for optimization. Fuel flow was the input signal and output response was pressure at the piston (IMEP).

BENEFITS

- Both objectives of the project, i.e., to apply the Taguchi/robust design methods to the idle quality problems and to improve the idle quality for the chosen application were accomplished.
- A 2.93-dB gain in S/N ratio was achieved, which was enough to produce a noticeable improvement in engine idle quality application.
- Improvement in the traditional quality indicator LNV (lowest normalized value) was 12%.
- A 14% increase in the efficiency was achieved, which resulted in an 8% increase in the idle fuel economy.

- The team gained valuable understanding and experience in robust design and its applications.
- The new design was robust against the engine mileage.
- This method is much superior to traditional measurements such NVH and LNV in terms of effectiveness.
- A new approach was found to evaluate the design so that product development time can be shortened significantly.

CASE STUDY

INTRODUCTION

The inherent random variation of engine combustion from one engine cycle to another results in indicated mean effective pressure (IMEP) variations. These variations in IMEP cause a deviation in the normal motion of the engine about its axis, resulting in engine vibration. This vibration is then transmitted to the vehicle body through the engine mounts, and ultimately felt as disturbances by the vehicle occupants. When the magnitude of the vibration is high, poor idle quality results. With increasing quality demands by the customers, the idle quality issue is of major concern.

Idle quality improvement can be achieved by improving engine combustion quality at idle and by better isolating the vehicle body from the engine, thereby reducing the level of vibration transferred from the engine to the vehicle body. Although it is necessary to optimize the engine and the frame together, only the engine idle quality will be discussed in this paper. The vehicle isolation approach is a subject of a separate study and will not be discussed further.

Several parameters affect engine idle combustion quality. Some of these parameters are idle speed, ignition timing, engine mileage, length of time at idle, and exhaust gas recirculation (EGR). While factors such as idle speed, EGR and ignition timing can be controlled (although this is sometimes expensive to do), other factors such as engine mileage and time at idle cannot be controlled by the engine designer. Therefore, some compensatory actions are taken during design in order to compensate for the effect of these uncontrollable factors. These actions usually result in inefficiencies and minimal benefits at best.

It is therefore more effective to find a solution that will improve idle quality by minimizing the effect of the uncontrollable factors in the field. Taguchi robust design methods have been successfully used on various quality engineering problems. When successfully applied, this method produces a system which is less sensitive to the uncontrollable factors. This method has been utilized to study some of the parameters affecting engine idle quality, and the findings are presented in this paper.

The Idle Combustion Quality Evaluation Process

A commonly used process for evaluating engine idle quality involves running the engine at idle for several cycles and collecting basic combustion data such as bum rates, cylinder pressures, air-fuel ratio (A/F) and actual ignition timing. Other parameters such as IMEP are then calculated from the raw data. Mean value, standard deviation, coefficient of variation and lowest normalized value (LNV) of IMEP are then calculated for the whole test. These statistical values are used to quantify the quality of the idle combustion.

"1-FAT" Design Method

The "1-factor-at-a-time" (1-FAT) method is used in most of the engine idle quality evaluation and development work done today. In the 1-FAT design method, parameters are varied one at a time while holding all other factors constant (or at their starting level). This is a time-consuming method where all the information obtained during the experiment is not fully utilized. Although this method is adequate for evaluating relative changes due to a particular parameter, a robust design is usually not obtained. Furthermore, it is an expensive process (when compared to robust design methods) since more tests are required to evaluate the effects of the same number of parameters.

The Robust Design Method

In the robust design method, several process parameters are evaluated at once using the minimum number of tests. The system's ideal function is optimized for best performance and robustness over a wider operating range. A robustness case study for idle combustion quality was initiated not only to demonstrate that Taguchi design methods can be successfully applied to this nontraditional problem, but to also achieve a robust engine idle operation. The challenge was to obtain a reasonably representative ideal function for the system. The control factors and noise factors were determined from a list of parameters (factors) based on past experiences with the system. An L_{18} orthogonal design was chosen based on the number of control factors, experimental setup and time, and cost.

Experiment Setup

Four load points representing the typical loads encountered during idle operation were considered for each of the 18 runs of the L_{18} array. Three samples were taken at each of the four load points for each test run for a total of 12 data sets for each run. The whole L_{18} array was run once for the low mileage engine and then repeated for the high-edge engine as required for the noise factor. A unique aspect of this experiment is that the input signal fuel flow was not set directly. Instead, the load points were set, and the input signal was then observed along with the system response.

Ideal Function

From the viewpoint of ideal function, IMEP is an approximately linear function of fuel flow at idle when air-fuel ratio (A/F), ignition timing and some of the engine combustion parameters are held constant under steady-state operation. When A/F is constant, air parameters are held constant under a flow that increases proportionately with fuel flow; therefore the load increases proportionately.

The ideal function chosen for this system is as in Figure 1.

Figure 1: Ideal Function

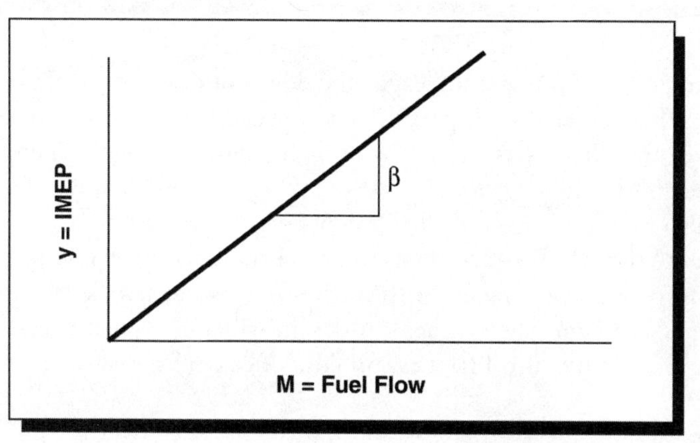

The ideal system response IMEP is shown to be a linear function of the input signal fuel flow (when the other parameters are held constant) in the required region. The goal of the experiment is now to maximize both the signal-to-noise ratio (η) and the slope (β). Needless to say, maximizing the S/N ratio is more important. The ideal function in this study has the following form:

$$Y = \beta M$$

where

$$y = \text{IMEP}; \quad \mathbf{M} = \text{Fuel Flow}; \quad \boldsymbol{\beta} = \text{Slope}$$

Control and Noise Factors

The control factors chosen for this study are shown in Table 1 and they include hardware and software parameters. Each of the control factors has three levels except for the cylinder head and the spark plug penetration which have two levels.

Table 1: Control Factors and Levels

Control Factors		# of Levels	Level 1	Level 2	Level 3
Cylinder Head, CH	A	2	Base	HISW	---
Fuel Injectors, INJ	B	3	Type I	Type II	Type III
Spark Plug Cap, SPG	C	3	Narrow	Base	Wide
Ignition Timing, IGT	D	3	Retarded	Base	Advanced
Air-Fuel Ratio, AFR	E	3	Rich	Stoich.	Lean
Spark Plug Penetration, SPP	F	2	Base	Long	---
Engine Spool, ES	G	3	Low	Medium	High
Injection Timing, IJT	H	3	Advanced	Base	Retarded

As an engine accumulates mileage, the idle combustion quality tends to degrade. This degradation is partly due to lower friction, valve deposits, worn parts and other factors associated with high engine mileage. Therefore, any good design should be robust against mileage accumulation. The design should also be robust against idle combustion quality variations over engines of the same model. For this experiment, all the uncontrollable factors were lumped into one noise factor with engine mileage at two levels, using low mileage and high mileage engines. A design which is robust over these two levels for the given control factors should reduce engine-to-engine variations.

Test Results

Signal-to-noise ratio and slope were calculated using all 24 data sets (12 for low and 12 for high mileage) for each of the 18 runs. The result of S/N ratios and β is shown along with the L_{18} matrix in Table 2. The overall mean value $\overline{\mathbf{T}}$ of the signal-to-noise ratios is calculated with the help of the following equation.

$$\overline{T} = \frac{1}{18}\sum_{i=1}^{18}\eta_i = 7.85 \text{ dB}$$

Analysis of means was then performed on the data of Table 2 to get the factor effects shown in Tables 2 and 3.

Table 2: Test Results

Test #	Control Factors								Response	
	CH	INJ	SPG	IGT	AFR	SPP	ES	IJT	S/N	β
1	Base	Type I	Narrow	Ret'd	Rich	Base	Low	Adv	8.39	9.67
2	Base	Type I	Base	Base	Stoich	Long	Med	Base	8.91	10.18
3	Base	Type I	Wide	Adv	Lean	Base	High	Ret'd	9.20	10.45
4	Base	Type II	Narrow	Ret'd	Stoich	Long	High	Ret'd	5.67	8.04
5	Base	Type II	Base	Base	Lean	Base	Low	Adv	8.65	10.87
6	Base	Type II	Wide	Adv	Rich	Base	Med	Base	11.67	10.95
7	Base	Type III	Narrow	Base	Rich	Base	Med	Ret'd	8.24	9.52
8	Base	Type III	Base	Adv	Stoich	Base	High	Adv	9.53	10.19
9	Base	Type III	Wide	Ret'd	Lean	Long	Low	Base	7.18	9.71
10	HISW	Type I	Narrow	Adv	Lean	Long	Med	Adv	11.96	11.66
11	HISW	Type I	Base	Ret'd	Rich	Base	High	Base	6.47	7.94
12	HISW	Type I	Wide	Base	Stoich	Base	Low	Ret'd	10.10	11.19
13	HISW	Type II	Narrow	Base	Lean	Base	High	Base	6.77	9.00
14	HISW	Type II	Base	Adv	Rich	Long	Low	Ret'd	13.33	12.54
15	HISW	Type II	Wide	Ret'd	Stoich	Base	Med	Adv	6.99	8.82
16	HISW	Type III	Narrow	Adv	Stoich	Base	Low	Base	11.27	12.15
17	HISW	Type III	Base	Ret'd	Lean	Base	Med	Ret'd	5.81	8.64
18	HISW	Type III	Wide	Base	Rich	Long	High	Adv	8.68	8.96

Table 3: Control Factor Effects

a) Signal-To-Noise Ratio (η)

Control Factors	CH	INJ	SPG	IGT	AFR	SPP	ES	IJT
Level 1	7.63	8.20	7.75	5.78	8.49	7.62	8.85	8.06
Level 2	8.07	7.88	7.81	7.59	7.78	8.32	7.96	7.74
Level 3		7.48	8.00	10.19	7.29		6.75	7.75

b) Slope (β)

Control Factors	CH	INJ	SPG	IGT	AFR	SPP	ES	IJT
Level 1	9.95	10.18	10.01	8.81	9.93	9.95	11.02	10.03
Level 2	10.10	10.04	10.06	9.95	10.10	10.18	9.96	9.99
Level 3		9.86	10.01	11.32	10.06		9.10	10.07

Confirmation Test Results

The optimum configuration for the system under study consists of factor levels that maximize the overall signal-to-noise ratio. Table 4 shows both the starting and optimal levels for the system. Using these factor levels, estimates of both the optimum and starting S/N can be obtained. Results of confirmation test runs to verify these estimates are also presented.

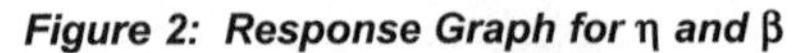

Figure 2: Response Graph for η and β

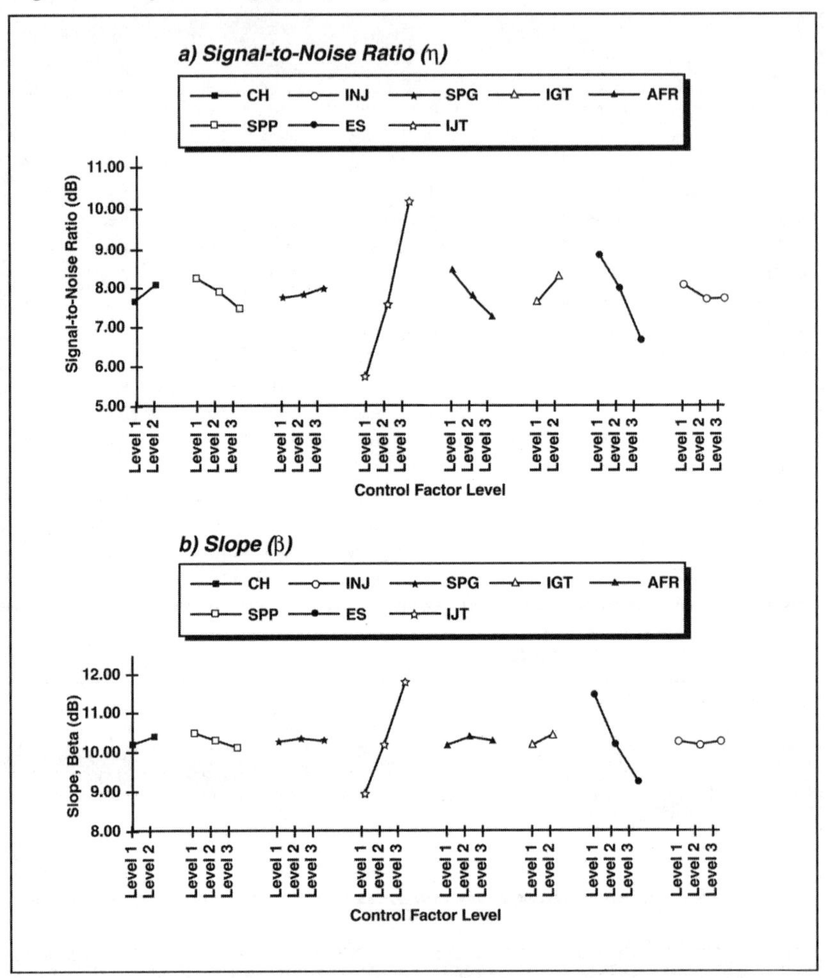

Using Tables 3 and 4, the starting and optimum signal-to-noise ratios were predicted. The predicted gain in S/N is 12.77 dB – 7.93 dB = 4.84 dB.

Table 4: Starting Design vs. Optimum Design

	CH	INJ	SPG	IGT	AFR	SPP	ES	IJT
Config.	A	B	C	D	E	F	G	H
Starting	Base	Type II	Base	Base	Stoich	Base	Low	Base
Optimum	Base	Type I	Wide	Adv.	Rich	Long	Low	Adv.

The actual S/N calculated from actual confirmation test runs are as follows:

$$\eta \text{ starting} = \overline{A_1} + \overline{B_2} + \overline{C_2} + \overline{D_2} + \overline{E_2} + \overline{F_1} + \overline{G_1} + \overline{H_2} - 7\overline{T} = 9.65 \text{ dB}$$

$$\eta \text{ optimum} = \overline{A_1} + \overline{B_1} + \overline{C_3} + \overline{D_3} + \overline{E_1} + \overline{F_2} + \overline{G_1} + \overline{H_1} - 7\overline{T} = 12.58 \text{ dB}$$

Therefore, the actual gain in S/N is 2.93 dB.

Predicted and actual values for slope (β) were calculated for the original and optimum configurations using equations similar to those used for S/N above. The results are shown in Table 5.

In addition to the 2.93-dB gain in S/N, there was an 12% approximate improvement in the engine idle quality indicator, LNV. This compares to the predicted value of approximately 20 %. The 14% increase in β corresponds to an 8% increase in idle fuel economy.

Table 5

Design	S/N Ratio	Beta Value
Current	9.65dB	11.12%
Optimal	12.58dB	12.67%
Gain	2.93dB	13.9%

Conclusion

The objectives for this project were to apply Taguchi robust design methods to the idle quality problem, and to improve idle quality for the chosen application. Both of these objectives were accomplished. An optimum configuration was found (Table 5) with a significant gain in signal-to-noise ratio. This gain was enough to produce a noticeable improvement in engine idle quality on this application. This improvement in idle quality should result in improved customer satisfaction associated with idle quality issues, and better product for the customers. It should be noted that a new approach was found to evaluate their design so that product development time can be shortened significantly. In addition, the authors gained valuable understanding and experience in robust design and its applications.

This case study is contributed by Waheed D. Alashe, Ellen B. Barnes, Kenneth L. Bath, Kenneth E. Jacobsen and Mary K. O'Keefe of Ford Motor Company, USA. The authors would like to thank John King – Manager at Ford Design Institute – for his special guidance throughout this study and would also like to thank the members of Ford's Powertrain System Analysis Department for their help and cooperation in collecting all the data used in this study.

CHAPTER SEVEN

Fuel Delivery System Robustness

FORD MOTOR COMPANY – USA

EXECUTIVE SUMMARY

BACKGROUND

Excessive fuel returned from the combustion area back to the gas tank is no longer allowed because of air pollution. This phenomenon is called hot fuel handling. Because an excess of fuel is not allowed, there is a strong need to develop a fuel pump system which delivers accurate fuel flow.

Fuel flow is controlled by electronics. Electronic control is the future of the engineered system. For any manufacturing company, it is important to establish an effective product development strategy to develop a robust electronic control system.

The project team had three years to develop the new system. Since they progressed much even after one and half years, the project was started.

PROCESS

Instead of evaluating the fuel delivery system by testing it to find problems, a robust design approach was used. This approach evaluates the design by measuring the variability of the energy transformation of the system. It looks for design that has the least variability of energy transformation regardless of a customer's usage.

Instead of repeating the design-build-test cycle several times until the design is good enough, the robust design approach systematically optimizes the design by studying several design parameters simultaneously. The team spent most of the energy to discuss what should be measured and how it should be tested. It took several four-hour meetings to plan the testing.

Five design parameters were tested under two noise conditions which were generated by varying fuel type, fuel temperature and tank pressure, and by indicating variability in customer usage conditions.

Benefits

- The team was able to complete the task several months prior to schedule.
- The new design is robust not only against customers' usage conditions but also against back pressure, which was achieved by developing a robust compensation against change in back pressure.
- No phone calls from the proving ground in Arizona complaining about problems regarding the fuel delivery system were received. No redesign was necessary.
- Technical knowledge developed by this study will be applicable to future products.
- At least a 20% reduction in field returns using the optimal fuel delivery system is forecasted.

CASE STUDY

Introduction

Hot Fuel Handling Robustness

Today's vehicles require the fuel system performance to be consistent and predictable over the operating range regardless of environmental conditions. This study provides a method by which Ford Motor Company identified the components that yield a robust fuel delivery system.

Inconsistent liquid fuel at the rail may cause customer dissatisfaction manifested in the following vehicle symptoms:

- Difficult restart after engine-off soak
- Rough or rolling engine idles
- Engine stumbling while cruising, accelerating and decelerating

Insufficient liquid fuel at the fuel injector is a root cause for these symptoms experienced by customers in the field. Higher temperatures in the engine, underbody, and fuel tank as well as high volatility fuel lead to fuel vaporization which may cause insufficient fuel delivery to the injectors. The higher temperatures and fuel characteristics cannot be controlled or specified by engineering so they are considered noise factors. A robust system or subsystem must perform consistently regardless of the noise factors to which it is exposed.

This study examines the fuel delivery subsystem to specify the fuel delivery components that contribute to a robust fuel system.

Fuel System Overview

Ford Motor Company continually strives for improved quality and new technologies to increase customer satisfaction. A new fuel system being developed targets increased fuel system robustness in the varied environmental conditions to which the vehicle is exposed. Brief overviews of the existing conventional fuel system and the new fuel system follow.

Conventional Fuel System

The conventional system applies battery voltage to the pump and regulates the fuel rail differential pressure via a mechanical regulator. The fuel not consumed by the engine is returned to the tank. The typical operating range for the fuel pump is 12 to 14 volts and 200 to 300 kPa.

New Fuel System

Ford Motor Company is developing a fuel system which regulates the pump speed and pressure. The speed and pressure range of the fuel pump is greater with this new fuel system than with the conventional system used in all Ford cars and light trucks. The typical operating range for the fuel pump is 6 to 13 volts and 200 to 400 kPa. The fuel pressure in the new system is maintained through an electronic controller via closed-loop feedback. The controller contains a table of the fuel pump performance with flow as a function of pressure and voltage. When the actual pressure differs from that desired, the controller adjusts the pump voltage to compensate for the pressure difference.

The pump flow as a function of voltage and pressure must be accurate throughout the operating range and under all conditions to minimize the error for which the electronic controller must compensate.

EXPERIMENT DESCRIPTION

Test Method

Parameter design using Taguchi's design of experiment method uses a system that aspires to an ideal function of performance. The ideal function is the measured response to a system input signal. All external and environmental factors are considered noise. Comparing system performance when subjected to favorable and unfavorable noises allows assessment of system robustness. A robust system will minimize the performance differences when subjected to the various noises.

Ideal Function

The ideal function for this study is based on physics. It is derived from fuel pump efficiency. Pump voltage was varied to change the power supplied to the fuel pump. Fuel pump current was observed and multiplied by the pump voltage to represent the system input signal. Flow rate was observed as the output response at four back pressure levels and five voltage levels. Figure 1 contains a schematic of the engineered system and the ideal function.

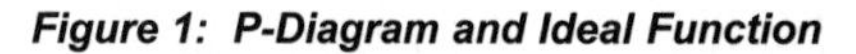

Figure 1: P-Diagram and Ideal Function

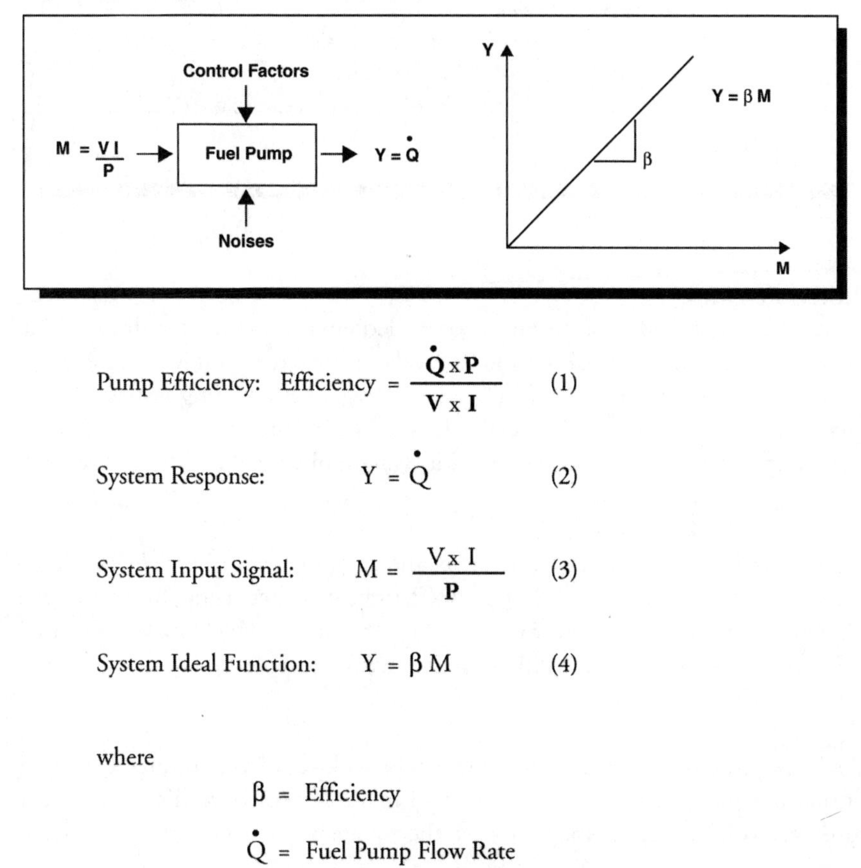

Pump Efficiency: $\text{Efficiency} = \dfrac{\dot{Q} \times P}{V \times I}$ (1)

System Response: $Y = \dot{Q}$ (2)

System Input Signal: $M = \dfrac{V \times I}{P}$ (3)

System Ideal Function: $Y = \beta M$ (4)

where

β = Efficiency

$\dot{Q}$ = Fuel Pump Flow Rate

P = Fuel System Back Pressure

V = Fuel Pump Voltage

I = Fuel Pump Current

The levels of the input signals V and P are in Tables 1 and 2. The fuel pump in the new fuel system is exposed to the ranges of voltages and pressures listed.

Table 1: Fuel Pump Voltage Levels

Pump Voltage (volts)	V_1	V_2	V_3	V_4	V_5
	6	8	10	12	14

Table 2: Fuel Pump Pressure Levels

Test Pressure (kPa)	P_1	P_2	P_3	P_4
	200	250	300	350

Noise Factors

As mentioned above, noise is an uncontrolled environmental condition. The noise factors and their compounded levels are listed in Table 3. The noise factors were selected to specifically address hot fuel handling issues. Noise factors level 1 are unfavorable to the fuel delivery system; they tend to lower the pump's flow. Noise factors level 2 are favorable: they tend to increase the pump's flow.

Fuel volatility (RVP values) represents the range of commercially available fuels in summer months. Higher RVP indicates increasing fuel volatility. Higher volatility fuels allow more vapor generation, which makes the fuel delivery system more susceptible to being unable to provide liquid fuel to the fuel injectors.

Fuel temperature values are the range of in-tank fuel temperatures expected during the summer months. Higher fuel temperatures have the same effect as higher volatility fuels; they increase the fuel vaporization.

Table 3: Noise Factors and Levels

Noise Factors	Fuel Type (Reid Vapor Pressure, RVP)	Fuel Temperature (°C)	Tank Vapor Pressure (in. H_2O)
N_1 (low pump flow)	High	High	0
N_2 (high pump flow)	Low	Low	10

The tank vapor pressure range is based on government regulations. Higher tank vapor pressures will have the tendency to increase fuel pump flow since there is an additional static fluid head at the fuel pump inlet.

Control Factors

Unlike noise factors, engineering can influence control factors. The control factors chosen for this experiment are specified or recommended by fuel delivery engineering as shown in Table 4.

Table 4: Control Factors and Levels

Control Factors	Level 1	Level 2	Level 3
A. Fuel Pump Type	Turbine	Gerotor	N/A
B. Assembly Type	Fuel Delivery Module (FDM)	Bracket	Bracket with Jet Pump
C. Mounting Angle (from vertical)	0	45	80
D. Rated Pump Flow (Lph)	Low	Medium	High
E. Modulation Frequency (kHz)	4	9.6	19.2

Fixed Factors

The following fuel system factors are fixed:

- Fuel level in tank
- Fuel heating rate
- Fuel tank size
- Fuel filter sock and its orientation to fuel in the tank

Experiment Test Results

Table 5 summarizes the results from L_{18}.

Since our goal is to minimize variations when exposed to the different noise factor levels, we are primarily concerned with maximizing the S/N ratio. The effects of the control factors on S/N are shown in Figure 2 and summarized in Table 6. A sensitivity analysis (analysis on β) discussion follows.

Table 5: Experiment Design and Results

Test	A	B	C	D	E	S/N	Beta
1	1	1	1	1	1	25.42	415
2	1	1	2	2	2	23.04	260
3	1	1	3	3	3	22.90	400
4	1	2	1	1	2	28.21	493
5	1	2	2	2	3	22.04	387
6	1	2	3	3	1	22.57	423
7	1	3	1	2	1	17.10	255
8	1	3	2	3	2	20.55	391
9	1	3	3	1	3	23.79	352
10	2	1	1	3	3	20.19	348
11	2	1	2	1	1	17.48	222
12	2	1	3	2	2	22.22	384
13	2	2	1	2	3	19.59	363
14	2	2	2	3	1	14.20	316
15	2	2	3	1	2	18.61	346
16	2	3	1	3	2	22.17	430
17	2	3	2	1	3	17.10	255
18	2	3	3	2	1	20.62	366

Table 6: Control Factors

	Control Factor A	Control Factor B	Control Factor C	Control Factor D	Control Factor E
Level 1	22.85	21.88	22.11	21.80	19.57
Level 2	19.15	20.90	19.07	20.77	22.50
Level 3	------	20.22	21.82	20.43	20.94

Figure 2: S/N Responses

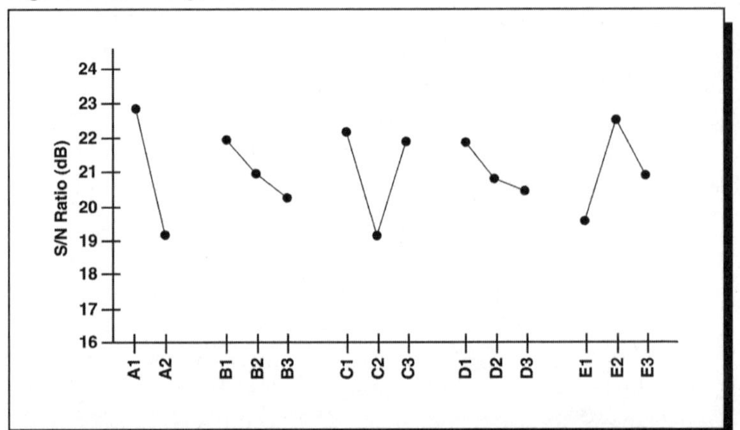

SENSITIVITY (β) ANALYSIS

Our customer is more concerned with predictable flow to the engine than with the overall efficiency of the fuel delivery system. For that reason, the sensitivity analysis is secondary in this study. The differences between the optimal system for maximizing S/N ratio and the system maximizing β are control factors B and D. Figure 3 and Table 7 show the control factor level results for sensitivity.

Figure 3: Sensitivity β Analysis

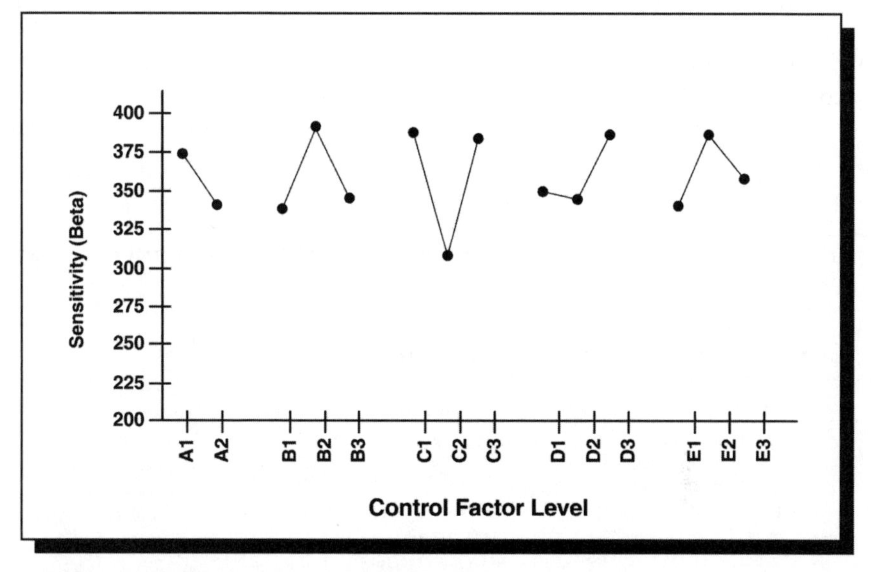

Table 7: Control Factor Results

	Control Factor A	Control Factor B	Control Factor C	Control Factor D	Control Factor E
Level 1	375	338	387	347	338
Level 2	341	391	305	343	384
Level 3	------	345	382	385	354

Our system response is the fuel flow to the engine. Consistent fuel flow to the engine is our customer's expectation from the fuel delivery system. Two of the fuel delivery assembly control factors divert fuel pump outlet fuel flow from the engine into the fuel tank. This will decrease the efficiency of the system. Since β is proportional to pump efficiency, β decreases with the systems that divert the flow to the tank. This is evident in control factor B. Level B2 is considerably higher than B1 or B3.

$\overline{T}$ = Overall Average of S/N = 21.0

The optimal fuel delivery system to maximize S/N is:

Pump Type:	Turbine
Assembly Type:	FDM
Mounting Angle:	00 from vertical
Pump Flow:	Low
Modulation Frequency:	9.6 kHz

Confirmation Test Results

We performed confirmation and verification tests to validate the L_{18} experimental results. They are divided into two sections:

- Bench tests
- Vehicle tests

For each set of tests the initial design is compared to the optimal design as determined by the L_{18} orthogonal array.

The initial configuration fuel delivery system is:

Pump Type:	Generator
Assembly Type:	Bracket
Mounting Angle:	800 from vertical
Pump Flow:	Low
Modulation Frequency:	19.2 kHz

Bench Test Confirmation

The two system configurations were tested using the same test procedure and method.

Initial Fuel Delivery System

The predicted S/N for the initial fuel delivery system is 20.6 dB as calculated in Eq. (5).

$$\eta_{initial} = \overline{T} + (\overline{A}_2 - \overline{T}) + (\overline{B}_2 - \overline{T}) + (\overline{C}_3 - \overline{T}) + (\overline{D}_1 - \overline{T}) + (\overline{E}_3 - \overline{T})$$
$$= 20.6 \text{ dB} \qquad (5)$$

Optimal Fuel Delivery System

The predicted S/N for the most robust fuel delivery system is 27.1 dB as calculated in Eq. (6).

$$\eta_{optimal} = \bar{T} + (\bar{A}_1 - \bar{T}) + (\bar{B}_1 - \bar{T}) + (\bar{C}_1 - \bar{T}) + (\bar{D}_1 - \bar{T}) + (\bar{E}_2 - \bar{T}) = 27.1 \text{ dB} \qquad (6)$$

Prediction and the result from confirmation are summarized in Table 8 below.

The bench confirmation tests yielded a 9.1-dB increase from the original fuel delivery system to the optimal one.

The difference between the actual and predicted values for S/N in the confirmation test indicates the possibility of some interactions and/or some unknown noise factor(s).

Table 8: Prediction and Confirmation

	Prediction	Confirmation
Initial Design	20.6 dB	14.9 dB
Optimal Design	27.1 dB	24.0 dB
Gain	6.5 dB	9.1 dB

Vehicle Verification

Verification in the vehicle is the successful completion of a hot fuel handling test. The test outline is:

- Warm Up/Grade Load/Engine Soak
- City Drive
- Extended Idle/Engine Soak

The fuel delivery system is exposed to a condition similar to noise level 2 in the beginning of the test. As the test progresses the condition in the fuel tank changes toward noise level 1. That may cause insufficient liquid fuel at the fuel rail. When there is insufficient liquid fuel, the symptoms of a nonrobust fuel delivery system may occur.

Initial Fuel Delivery System

The initial system did not pass the hot fuel handling test. During the grade load section the engine stalled due to insufficient fuel pressure. Figure 4 contains a plot of the actual fuel pressure compared to the requested fuel pressure. Note the increasing difference between the requested and actual pressure. The actual pressure fluctuations are due to vapor handling difficulty at the fuel pump.

Figure 4: Initial Fuel Delivery System Performance

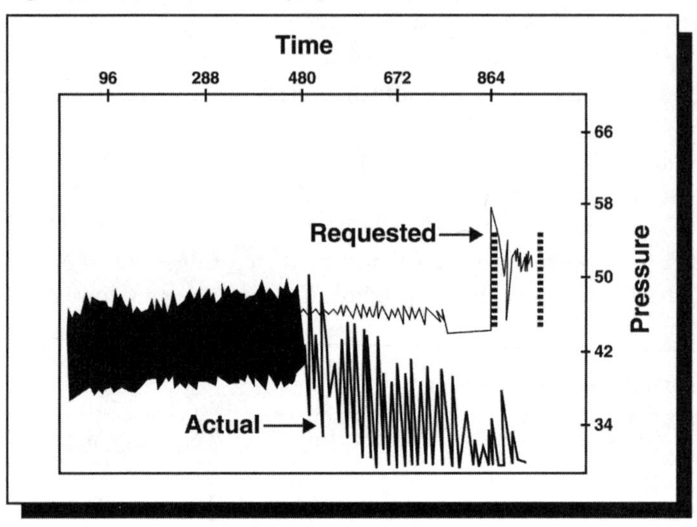

Optimal Fuel Delivery System

The optimal fuel delivery system successfully passed the hot fuel handling test. Figure 5 is the plot of actual fuel pressure and requested fuel pressure during the grade load part of the test. The requested and actual pressure signals overlap with the optimal system indicating that the optimal system provides the pressure required.

Figure 5: Optimal Fuel Delivery System Performance

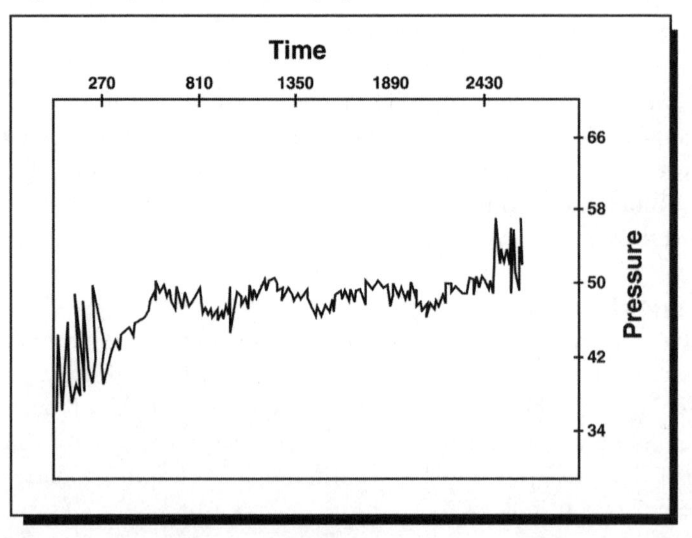

Conclusion

Parameter design for a robust delivery system indicates that the following parameters yield the optimal delivery system:

- Turbine Fuel Pump
- Fuel Delivery Module
- 00 from Vertical Orientation
- Lowest Pump Capacity
- 9.6-kHz Pump Modulation Frequency

The optimal design proved to have better performance than the initial design in both bench testing and vehicle evaluation.

The primary benefit associated with this approach to problem solving is increased customer satisfaction. Our internal Ford customers are more satisfied since they can continue vehicle development without concerns regarding the fuel delivery system. Our customers will also be more satisfied due to the decreased likelihood of hot fuel handling issues while they own a Ford vehicle. We estimate that 20% of our field returns will be avoided using the optimal fuel delivery system.

Other benefits are decreased development time and cost. An alternative approach used to address this issue consisted of multiple bench studies followed by vehicle studies. That approach resulted in the same conclusion; however, it took approximately six months longer due to learning in smaller incremental steps, and it cost an additional $25,000.

This case study is contributed by J.S. Colunga, D. Lau, J.R. Otterman, B.C. Prodin, K.W. Turner, and J.J. King, of Ford Motor Company, USA.

CHAPTER EIGHT 

Wiper System Chatter Reduction

FORD MOTOR COMPANY – USA

EXECUTIVE SUMMARY

BACKGROUND

The wiper arm/blade forms an integral part of the windshield wiper system. Chatter is the phenomenon that occurs when the wiper blade does not take the proper set and "skips" across the windshield during operation, potentially causing both acoustic and wipe quality deterioration, which significantly affect the satisfaction of the customers.

The major quality concern, which inspired the robustness study, was the wiper chatter phenomenon in some of the windshield wiper systems. Customer expectations the good windshield wiper systems are:

- Clear vision under a wide variety of operating conditions
- Multiple level of wiper speed
- Uniform wiper pattern
- Quiet under all weather conditions
- Long life and high reliability

PROCESS

Instead of measuring wiper quality deterioration, a robust design approach was used by evaluating the actual time that a blade takes to reach a fixed point on the windshield at a given cycle. That means that the focus was on measuring the functionality of the system and not the end characteristics.

Nine control factors were used under all categories of noise factors. Eighteen design configurations were tested in this study. It is important to note that the confirmation runs of the study proved that there were no severe interactions among control factors to variability. This clearly shows the power of measuring the functionality of the system. While Taguchi Methods™ emphasize the need of measuring functionality, the traditional DOE (design of experiment) focuses on measuring the symptoms (output characteristics).

This difference is not understood by many people and they tend to confuse Taguchi Methods with DOE. In other words Taguchi Methods focus on measuring the right things with fewer runs to obtain the best possible results. This clearly shows how much time, money and effort could be saved by using the Taguchi Methods. Since this approach emphasizes measurement of functionality, the knowledge level of an engineer using the process would be enhanced. The experiment was dynamic in nature because it chose an appropriate input signal with proper levels.

Benefits

- The robust study for a windshield wiper system indicates that the chlorination, graphite, arm rigidity and superstructure rigidity have significant impacts to the optimal wiper system.
- Load distribution on the blade has minimal influence.
- As a result, the low friction and high rigidity of the wiper arm blade will lead to a more robust windshield wiper system.
- The new wiper system is robust against the following noise categories: piece-to-piece variation, changes in dimensions, customer usage/duty cycle, external and internal environment. Because of this, it is expected to result in higher reliability and reduced warranty.
- The team gained knowledge in the operation of wiper applications and the application of robust design principles.

CASE STUDY

Introduction

The Windshield Wiper System

A windshield wiper system is composed of four major subsystems: (1) the wiper motor, (2) the wiper linkage assembly, (3) the pivot towers, and (4) the wiper arm/blade combination. The wiper motor provides the driving power to the system. The wiper linkage assembly (usually a series- or parallel-coupled four-bar mechanism) is a mechanical transmission configuration to transfer the driving power from the motor to the pivots and to change the rotary motion of the motor to an oscillating motion at the pivots. The pivot towers provide a firm anchor from which the pivots operate, holding the wiper pivots in the proper orientation with the windshield surface. When the arm/blade combinations are attached to the pivot and the system is cycled, the blades clear two arc-shaped wipe patterns on the windshield.

The customer expectations for a good windshield wiper system are:

- Clear vision under a wide variety of operating conditions
- Multiple level of wiper speed
- Uniform wiper pattern
- Quiet under all weather conditions
- Long life and high reliability

What Is Chatter?

Chatter is the phenomenon that occurs when the wiper blade does not take the proper set and "skips" across the windshield during operation, potentially causing both acoustic and wipe quality deterioration, which significantly affect the satisfaction of our customers.

The major quality concern, which inspired this robustness study, was the wiper chatter phenomenon in some of the windshield wiper systems. From the results of previous studies it was determined that the components which contribute most to chatter are the arms and blades. This study was conducted to more thoroughly understand the specific design factors relating to arm/blade combinations that affect system performance.

Description of the Experiment

The Engineered System

The control factors and the noise factors are selected based on the knowledge of chatter. The engineered system is shown in Figure 1.

Figure 1: The Engineered System

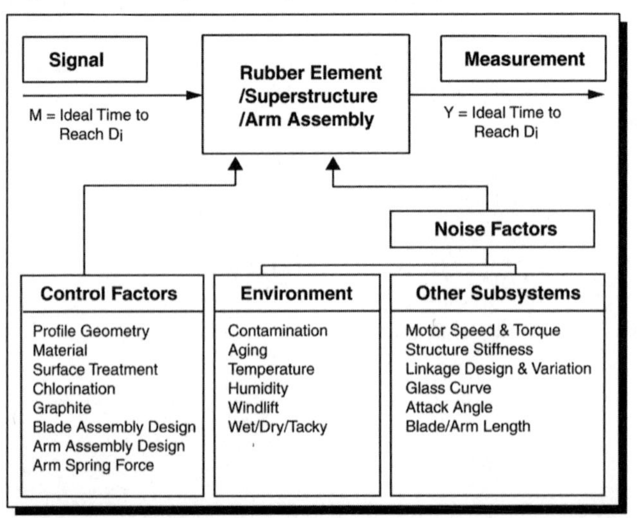

The Ideal Function

The ideal function for this study is based on the following hypothesis: for an ideal system, the actual time for the wiper blades to reach a fixed point on the windshield during a cycle should be the same as the theoretical time (ideal time) for which the system was designed. Under the presence of the noise factors, the actual time will differ from the theoretical time. Furthermore, the actual time of a robust system under varying noise conditions should have less variation, and be closer to the theoretical time. Therefore, the system ideal function is given by

$$\mathbf{Y_i = \beta M_i}$$

where

$\mathbf{Y_i}$ = Actual time that the blade reaches a fixed point on the windshield at the ith cycle.

$\mathbf{M_i}$ = Theoretical time for blade to reach a fixed point on the windshield at the ith cycle.

$$\beta = \frac{i}{\text{(RPM of the motor)}}$$

The system ideal function is shown in Figure 2.

Figure 2: The Ideal Function

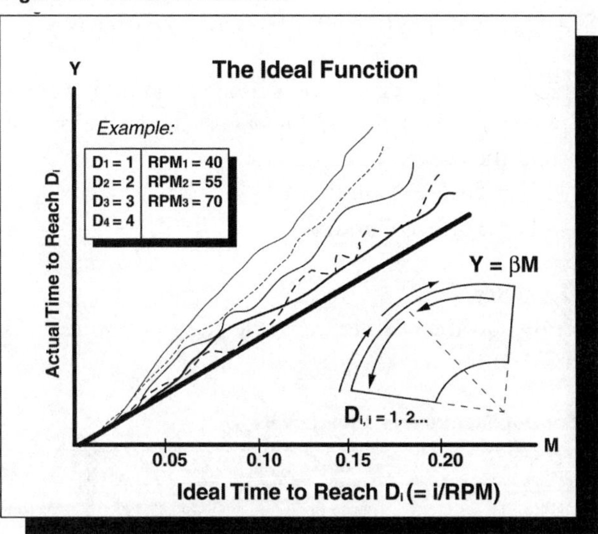

Three RPM values were used in the experiment: 40, 55 and 70.

In general, due to the noise effects, the actual time would be always longer than the theoretical time determined by the design, i.e., $\beta > 1$, and $\beta = 1$ would be the ideal case.

In this experiment, other measurements were also observed and those are:

- The lateral load on the wiper arm
- The normal load on the wiper arm

The Noise Strategy

In general, noises can be categorized in five categories:

1. Piece-to-piece variation – manufacture, material variation, etc.
2. Change in dimension due to wear, fatigue, etc.
3. Customer usage/duty cycle
4. External environment – climate, road, wind, temperature, etc.
5. Internal environment – interface with other body systems

In this experiment, the important noise factors were grouped into two (2) noise factors each with two levels:

1. Noise factor S included factors such as wiper angular orientation in park position, temperature, humidity and condition of blades, which would produce a tendency of AC variability, i.e., variation about the mean response time.
   $S_1 = -3^0$ at Park / 20^0C / 50% Humidity/ Before Aging
   $S_2 = 1^0$ at Park / 2^0C / 90% Humidity / After Aging

2. Noise factor T is the windshield surface condition, wet or dry, which would produce a tendency of DC variability in the response time, i.e., shifting the mean.
   T_1 = Wet Surface Condition
   T_2 = Dry Surface Condition

The Control Factors

The control factors and their levels were selected through team brainstorming as shown in Table 1.

Table 1: The Control Factors and Their Levels

	Factor	Level 1	Level 2	Level 3
A	Arm Lateral Rigidity	Design #1	Design #2	Design #3
B	Superstructure Rigidity	Median	High	Low
C	Vertebra Shape	Straight	Concave	Convex
D	Spring Force	Low	High	–
E	Profile Geometry	Geometry 1	Geometry 2	New Design
F	Rubber Material	Material I	Material II	Material III
G	Graphite	Current	Higher Graphite	None
H	Chlorination	Median	High	Low
I	Attach Method	Current Clip	Modified	Twin Screw

EXPERIMENT RESULTS AND DATA ANALYSIS

Design of Experiment

The L_{18} orthogonal array was selected for the DOE shown in Table 2. It may be seen that under each run, there are 12 tests under different RPM and the combined noise conditions.

Table 2: Design of Experiment

Run #	D	A	B	C	E	F	G	H	I	RPM 1				RPM 2				RPM 3			
										S1		S2		S1		S2		S1		S2	
										Wet	Dry	Wet	Dry	Wet	Dry	Wet	Dry	Wet	Dry	Wet	Dry
1	1	1	1	1	1	1	1	1	1												
2	1	1	2	2	2	2	2	2	1												
3	1	1	3	3	3	3	3	3	1												
4	1	2	1	1	2	2	3	3	2												
5	1	2	2	2	3	3	1	1	2												
6	1	2	3	3	1	1	2	2	2												
7	1	3	1	2	1	3	2	3	3												
8	1	3	2	3	2	1	3	1	3												
9	1	3	3	1	3	2	1	2	3												
10	2	1	2	3	3	2	2	1	2												
11	2	1	2	1	1	3	3	2	2												
12	2	1	3	2	2	1	1	3	2												
13	2	2	1	2	3	1	3	2	3												
14	2	2	2	3	1	2	1	3	3												
15	2	2	3	1	2	3	2	1	3												
16	2	3	1	3	2	3	1	2	1												
17	2	3	2	1	3	1	2	3	1												
18	2	3	3	2	1	2	3	1	1												

Test Setup

A special test fixture was built for this experiment, and three sensors were attached to the windshield to record a signal when a wiper blade passes them. Strain gauges were attached to the wiper arm to continuously record the lateral and normal loads to the wiper arm. Typical testing results are shown in Figure 3.

Figure 3: A Plot of Typical Testing Data

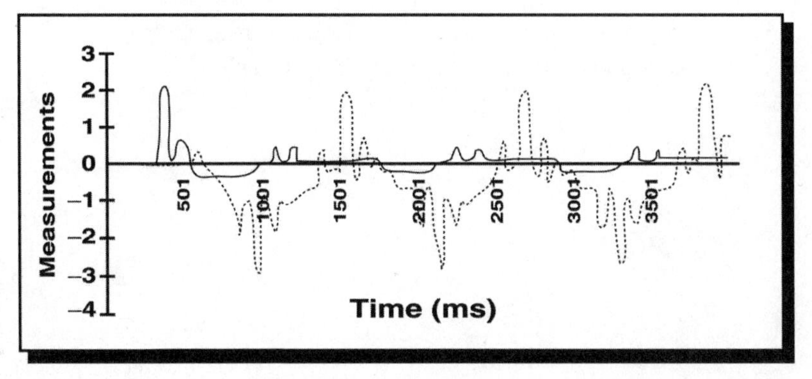

Test Results

In this paper, only the analysis of the time measurement data is presented. Under each test, the system was run for a certain period of operating time, and the actual accumulated operating time for the blade to pass a fixed point were recorded against the ideal operating time which is determined by the nominal RPM setting of the constant RPM motor. A testing data plot for this measurement is shown in Figure 4.

Figure 4: The Typical Testing Data Plot

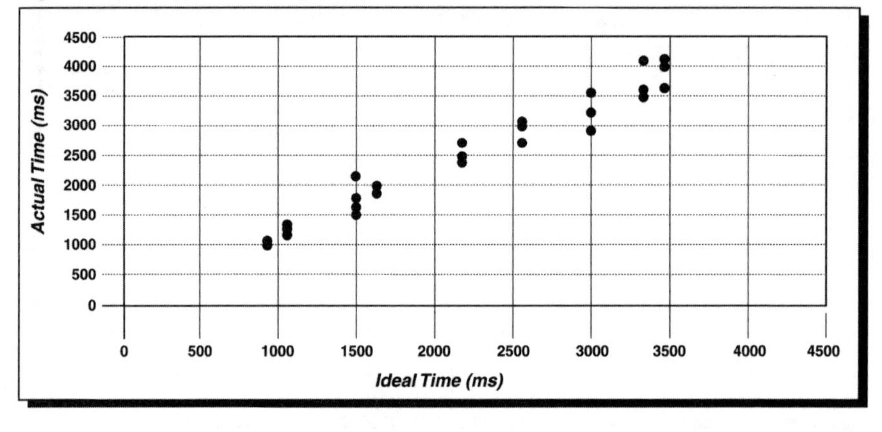

The regression analyses were conducted for those data and the slopes of the regression lines were estimated, and the mean square was used to estimate the variation.

In Table 3, the S/N ratio and β values are evaluated by

$$\text{S/N Ratio} = 10 \times \log_{10}\left(\frac{\beta^2}{\sigma^2}\right)$$

where **β** = Slope of regression line (forced through zero)
σ = Square root of mean squared deviation

Table 3: S/N Ratio and Beta Value for Each Run

Run #	D	A	B	C	E	F	G	H	I	S/N	Beta
1	Low	Design 1	Median	Straight	Geo 1	Material I	Current	Median	Current	–34.27	1.078
2	Low	Design 1	High	Concave	Geo 2	Material II	Higher	High	Current	–34.41	1.089
3	Low	Design 2	Low	Convex	New	Material III	None	Low	Current	–44.27	1.133
4	Low	Design 2	Median	Straight	Geo 2	Material II	None	Low	Modified	–37.27	1.096
5	Low	Design 3	High	Concave	New	Material III	Current	Median	Modified	–30.81	1.073
6	Low	Design 3	Low	Convex	Geo 1	Material I	Higher	High	Modified	–32.80	1.093
7	Low	Design 3	Median	Concave	Geo 1	Material III	Higher	Low	Twin Scr	–34.65	1.076
8	Low	Design 3	High	Convex	Geo 2	Material I	None	Median	Twin Scr	–38.89	1.094
9	Low	Design 3	Low	Straight	New	Material II	Current	High	Twn Scr	–33.39	1.065
10	High	Design 1	Median	Convex	New	Material II	Higher	Median	Modified	–30.46	1.063
11	High	Design 1	High	Straight	Geo 1	Material III	None	High	Modified	–35.24	1.079
12	High	Design 1	Low	Concave	Geo 2	Material I	Current	Low	Modified	–42.59	1.096
13	High	Design 2	Median	Concave	New	Material I	None	High	Twin Scr	–30.54	1.075
14	High	Design 2	High	Convex	Geo 1	Material II	Current	Low	Twin Scr	–32.52	1.068
15	High	Design 2	Low	Straight	Geo 2	Material III	Higher	Median	Twin Scr	–32.70	1.065
16	High	Design 3	Median	Convex	Geo 2	Material III	Current	High	Current	–31.00	1.065
17	High	Design 3	High	Straight	New	Material I	Higher	Low	Current	–33.67	1.075
18	High	Design 3	Low	Concave	Geo 1	Material II	None	Median	Current	–34.18	1.079

Data Analysis and the Optimal Condition

The S/N ratios and beta values given in the last two columns of Table 3 were further analyzed to determine the significance of the control factors and to select the best levels.

S/N Ratio

The plot of the S/N ratio against the control factors and their levels are shown in Figure 5. From the plot, it can be seen that different levels of the arm rigidity, superstructure rigidity, graphite and chlorination yield quite different results whereas load distribution has a relatively small effect on S/N.

Figure 5: The S/N Ratio Chart

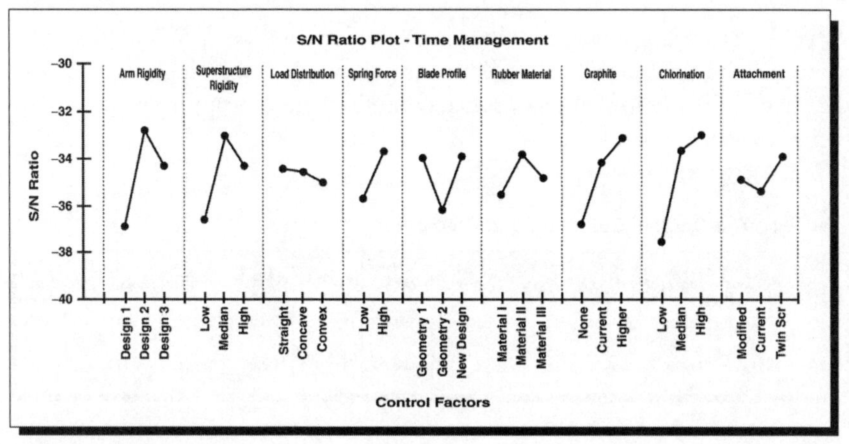

Beta Value

From Figure 6 it may be observed that in contrast to the plot of the S/N ratio, where the differences in the impact of the control factor levels on the S/N ratio are large, the difference in the impact of the control factor levels to the beta value are relatively small.

Figure 6: The Beta Value Chart

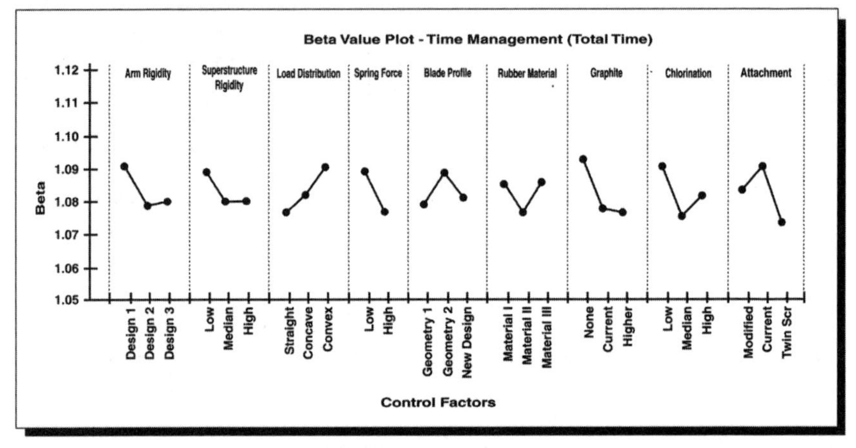

The Optimal Condition

The optimal condition was determined by resorting to two-step optimization. In two-step optimization, the selection of the control factor combination was carried out by (1) maximizing S/N ratio and (2) having a beta value which is close to 1.00.

Since the objective was to have minimum variability in motion and minimum difference between actual and theoretical times for the wiper blades to reach a fixed point on the windshield in each cycle, maximizing the S/N ratio was the top priority.

Table 4 gives the optimal condition and the baseline design. The predicted values of the S/N ratio and of β are also shown in Table 4. It may be seen that the S/N ratio of the optimal condition has an approximate 10 dB increase over the baseline.

Table 4: The Optimal Condition and Prediction

	Arm Rigidity	Super-structure	Load Distribution	Spring Force	Blade Profile	Rubber Material	Graphite	Chlorination	Attachment	S/N Ratio	Beta Value
Optimal	Design 2	Median	Straight	High	New Des	Material 2	Higher	Median	Twin Scr	-24.71	1.033
Baseline	Design 1	Median	Straight	Low	Geo 1	Material 1	Current	Median	Current	-35.13	1.082

The predicted values in Table 4 were calculated as follows:

S/N Ratio

S/N Ratio of Optimal $= -$ [34.647 + (32.77 – 34.647) + (33.03 – 34.647)
+ (34.42 – 34.647) + (33.66 – 34.647)
+ (33.86 – 34.647) + (33.70 – 34.647)
+ (33.11 – 34.647) + (33.55 – 34.647)
+ (33.78 – 34.647)] = –24.71

S/N Ratio of Baseline = - [(34.647 + (36.87 - 34.647) + (33.03 – 34.647)
+ (34.42 – 34.647) + (35.64 – 34.647)
+ (33.94 – 34.647) + (35.46 – 34.647)
+ (34.10 – 34.647) + (33.55 – 34.647)
+ (35.30 – 34.647)] = –35.13

Beta Value

Beta Value of Optimal = [1.08249 + (1.0783 – 1.08249) + (1.0792 – 1.08249)
+ (1.0763 – 1.08249) + (1.0765 – 1.08249)
+ (1.0807 – 1.08249) + (1.0767 – 1.08249)
+ (1.U768 – 1.08249) + (1.0752 – 1.08249)
+ (1.0737 – 1.08249)] = 1.033

Beta Value of Baseline = [1.08249 + (1.0898 – 1.08249) + (1.0792 – 1.08249)
+ (1.0763 – 1.08249) + (1.0885 – 1.08249)
+ (1.0788 – 1.08249) + (1.0852 – 1.08249)
+ (1.0780 – 1.08249) + (1.0752 – 1.08249)
+ (1.0904 – 1.08249)] = 1.082

THE CONFIRMATION TESTS AND RESULTS

The Confirmation Test Plan

Baseline Configuration: A_1 B_1 C_1 D_1 E_1 F_1 G_1 H_1 I_1
Optimal Configuration: A_2 B_1 C_1 D_2 E_3 F_2 G_2 H_2 I_3

Confirmation Test Results

	Predicted		Confirmed	
	S/N Ratio	Beta	S/N Ratio	Beta
Optimal	–24.71	1.033	–25.41	1.011
Baseline	–35.13	1.082	–36.81	1.01
Gain	10.42	----	11.4	----

CONCLUSION

This robust design study for a windshield wiper system indicates that the chlorination, graphite, arm rigidity and superstructure rigidity have significant impacts to the optimal wiper system. Load distribution on the blade has minimal influence. As a result, low friction and high rigidity of the wiper arm and blade will lead to a more robust windshield wiper system.

This case study is contributed by Michael Deng, Dingjun Li, and Wesley Szpunar of Ford Motor Company, USA. The authors wish to thank John King, John Coleman, Dr. Ken Morman, Henry Kopickpo, Mark Garaseia of Ford Motor Company; and Brian Thomas and Don Hutter for their support during the course of this work.

CHAPTER NINE 

Development of a Formula for Chemicals Used in Body Warmers

FUMAKILLA, LTD. – JAPAN

EXECUTIVE SUMMARY

BACKGROUND

Recently, the demand for body warmers has been rapidly increasing because of their convenience for use. But when the ambient temperature changes, the user may feel too warm or too cold. This is because the temperature generated by the body warmer varies depending on the environmental temperature. This study was aimed at developing a formula to maintain a constant temperature which is robust against environmental temperature changes. The heat-generating materials used for body warmers consist of iron powder, active carbon, water, salt and a water-retaining agent. These materials are packed in an air-permeable inner bag, which is further enclosed in an airtight outer bag.

The study was needed largely because of severe competition that has been forcing the market price down year after year. Therefore cost reduction has become an urgent necessity. Also, specifications for disposable bodywarmers were established in the JIS (Japanese Industrial Standards). Therefore, it is important to maintain quality by complying with the specifications in order to use the JIS mark.

PROCESS

In stead of resorting to a common one-factor-at-a-time approach, a more efficient robust design methodology was used and the functionality of the engineering system was studied. Seven control factors were tested with an orthogonal array L_{18} under two noise factors (exposing time and number of flannel sheets) representing the user's clothing variation. The ingredients were considered the input signal and temperature was the output response for this dynamic system.

BENEFITS

- A gain of 3.91 dB in the S/N ratio and a 34% increase in heat-generation efficiency were achieved without a cost increase.
- Because of the efficiency gain, it is now feasible to reduce the size and weight of body warmer while maintaining the target performance.

CASE STUDY

INTRODUCTION

The heat-generating materials used for body warmers consist of iron powder, active carbon, water, salt and a water-retaining agent. These materials are packed in an air-permeable inner bag, which is further enclosed in an airtight outer bag.

Recently, the demand for such body warmers has been rapidly increasing because of their convenience. But when the ambient temperature changes, the user may feel too warm or too cold. The user feels too warm when entering a heated room or feels too cold when going outside. This is because the temperature generated by the body warmer varies depending on the environmental temperature. This study, using quality engineering methods, attempts to develop a formula that maintains a constant temperature which is robust against environmental temperature changes.

As iron rusts, heat is generated. In the case of spontaneous combustion, heat generation cannot be perceived because of its slow reaction rate. Disposable body warmers are designed to utilize such heat by increasing its reaction speed. A body warmer consists of heat-generating substances such as iron powder, active carbon, water, salt and a water-retaining agent. These materials are packed in an inner bag with permeable holes for air to come in, then the inner bag is tightly sealed in an outer bag.

In an exothermic reaction, iron powder reacts with water and the oxygen in the air to become ferric hydroxide as shown in the following equation.

$$\mathbf{Fe + 3/4\ O_2 + 3/2\ H_2O \rightarrow Fe(OH)_3 + 96\ kcal}$$

The purpose of adding salt is to accelerate the reaction. Active carbon is used to absorb the oxygen in the air to increase the oxygen concentration. A water-retaining agent prevents the iron powder from becoming sludgy in the brine. The heat-generating time and temperature are adjusted by the permeability of the inner bag, the types of mixed materials, particle size and mixing ratios.

Objectives of the Study

The demand for disposable body warmers is increasing because of its simplicity and convenience. But severe competition has been forcing the market price down year after year, and cost reduction has become an urgent necessity. Also, specifications for disposable body warmers were established in the JIS (Japanese Industrial Standards). Therefore, it is important to maintain quality by complying with the specifications in order to use the JIS mark. As the demand has expanded, the body warmers have come to be used in various conditions.

When body warmers are used in the same environment for a long period of time such as in fishing, skiing or watching outdoor games, it is not difficult to maintain constant heat. But if a body warmer is started at home before going to the office, the environmental temperature changes when walking to a bus or a train station, in the bus or in the train where heaters are used and affects the temperature of the body warmers. Working in a heated office or going to a warehouse without heating, the user feels either too warm or too cold. This is because the temperature of the body warmer is affected by the ambient temperature. The purpose of this study is to find stable performance under different environmental conditions, together with another objective: cutting the cost down.

Ideal Function

A good body warmer must maintain a constant temperature for a certain target time duration without being affected by environmental temperature or clothing conditions. In this study, room temperature could not be changed since temperature was measured by the method stated in the JIS and is shown in Figures 1 and 2. But the number of flannel sheets covering the temperature measuring device was varied to simulate the amount of clothing worn by the user. Eight sheets are used to cover the equipment under the standard procedure. For another noise factor, a sample was started and exposed to the air for a short period of time before testing to simulate deterioration.

Control factors include the amount and concentration of ingredients such as iron powder, active carbon, a water-retaining agent, salt and water. Air permeability of the inner bag was also a control factor. The total amount of heat-generating substance and heat-maintaining time were considered signal factors.

The target heat-generating time was divided into four segments. The amount of heat generated in each of the four time segments are calculated as Y. Letting the heat-generated time segment and the total amount of ingredients

Figure 1: Heating Device

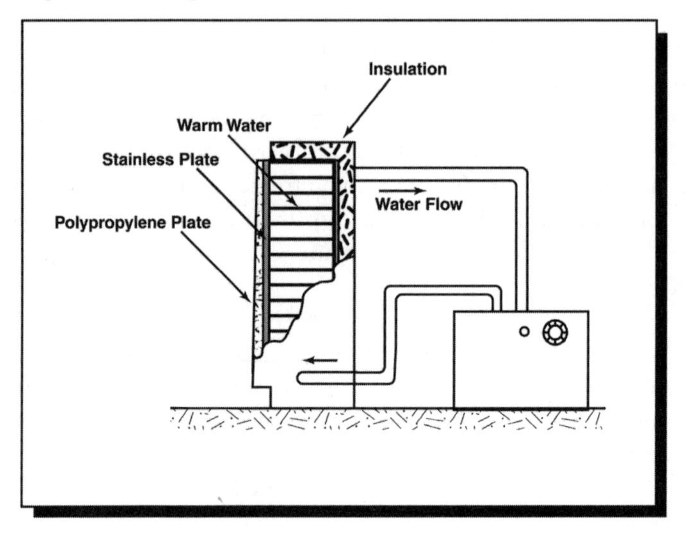

Figure 2: Sample Placement

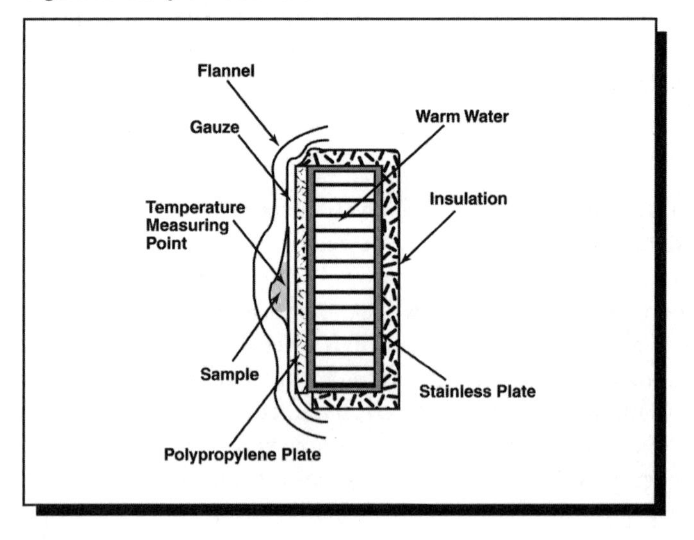

be M and M*, respectively, the following equation was used as the ideal function. Dynamic signal-to-noise ratio was used for evaluation.

$$Y = \beta MM^*$$

Factor Levels and Layout

Control factors, signal factors, and noise factors and their levels are shown in Tables 1, 2, and 3, respectively. The experimental layout is shown in Table 4.

Table 1: Control Factors and Levels

	Factor	Level 1	Level 2	Level 3
B	Water-Retaining Agent 1	–30%	Current	+30%
C	Water-Retaining Agent 2	–30%	Current	+30%
D	Active Carbon	–30%	Current	+30%
E	Concentration of Brine	–30%	Current	+30%
F	Amount of Brine	–30%	Current	+30%
G	Air Permeability	–30%	Current	+30%
H	Iron Powder	–30%	Current	+30%

Table 2: Signal Factors and Levels

M*	Amount of Ingredients
M*1	–30%
M*2	Standard
M*3	+30%

M	Target Heat-Generating Time
M1	1/4 of Target
M2	2/4 of Target
M3	3/4 of Target
M4	4/4 of Target

Table 3: Noise Factor and Levels

Level	Exposing Time	No. of Flannel Sheets
N1	30 min	1
N2	0 min	8

Table 4: Layout

No.	e	B	C	D	E	F	G	H	M*1		M*2		M*3	
									N1	N2	N1	N2	N1	N2
1	1	1	1	1	1	1	1	1						
2	1	1	2	2	2	2	2	2	(From each of 6 x 18 = 108 combinations, temperature was measured four times: 1/4, 2/4, 3/4, and 4/4 of target.)					
.														
.														
.														
18	2	3	3	2	1	2	3	1						

There are two water-retaining agents: wood powder which releases water during the earlier reaction period because of its low water-retaining capability, and a high polymer chemical with a high water-retaining capability for the later reaction period. These are denoted by water-retaining agents 1 and 2, respectively. Because the control factor levels change in each run of the orthogonal array, the total amount of ingredients varies. This is called the standard amount for each run. For the three levels of signal factor M' the standard amount was varied +/– 30%. There were four noise factor conditions.

1. Outer bag is sealed immediately.
2. Outer bag is sealed after inner bag was exposed to the air for 30 minutes.
3. Use one sheet of flannel cover.
4. Use eight sheets of flannel cover.

These four conditions were compounded into two noise levels. Control factors B through H were assigned to columns 2 through 8 of an L_{18}, respectively. In each experimental run, the total amount was considered the standard amount, then it was varied +/– 30% to generate three levels of input signal M*. With a two-level noise factor, six tests were conducted. In each test, the temperature was measured four times for each time segment. The data set from run No.1 of L_{18} is shown in Table 5. The calculation of signal-to-noise ratio and sensitivity from this data set is shown below Table 5.

Table 5: Results of Experiment No. 1

No.	M*	N	Average Temperature (°C)				Amount of Heat Generated				L	Weight
			1/4	2/4	3/4	4/4	Y 1	Y 2	Y 3	Y 4		
1	M*1	N1	52.1	49.1	45.8	43.2	92.82	160.44	199.08	221.76	222644.44	27.93
		N2	54.8	53.8	50.9	48.2	104.16	199.92	263.34	305.76	295266.24	27.93
	M*2	N1	47.5	45.3	42.2	39.7	105.00	183.60	219.60	232.80	493566.95	39.89
		N2	53.8	52.0	49.5	46.7	142.80	264.00	351.00	400.80	796284.18	39.89
	M*3	N1	49.5	47.8	44.6	41.5	152.10	277.68	341.64	358.80	1281311.45	51.86
		N2	51.2	50.7	48.6	45.5	165.36	322.92	435.24	483.60	1638791.35	51.86

Total variation S_T

$$S_T = 92.82^2 + 160.44^2 + \cdots + 435.24^2 + 483.60^2 = 1{,}758{,}524.77 \qquad (f = 24)$$

It must be noted that the linear equation, L, and the effective divider, $r_0 r$, are different from time to time because of the different values of M* and M.

$$r_0 r = 2 \times \{27.93^2 \times (4.2^2 + \cdots + 16.8^2) + 39.89^2 \times (6^2 + \cdots + 24^2 + 51.86^2 \times (7.8^2 + \cdots + 31.2^2)\} = 14{,}080{,}263.32$$

$$L(M^*_1 N_1) = (4.2 \times 92.82 + 8.4 \times 160.44 + 12.6 \times 199.08 + 16.8 \times 221.76) \times 27.93 = 222{,}644.44$$

Other linear equations are similarly calculated and the results are shown in Table 5.

$$L_1 = L(M^*_1N_1) + L(M^*_2N_1) + L(M^*_3M_1) = 1{,}997{,}522.84$$
$$L_2 = L(M^*_1N_2) + L(M^*_2N_2) + L(M^*_3M_2) = 2{,}730{,}341.77$$

Variation of the proportional tern, S_β

$$S_\beta = (L_1 + L_2)^2 / r_o r$$
$$= (1{,}997{,}522.84 + 2{,}730{,}341.77)^2 / 14{,}080{,}263.32 = 1{,}587{,}520.30 \qquad (f = 1)$$
$$S_{\beta \times N} = (L_2 - L_1)^2 / r_o r$$
$$= 38{,}140.17 \qquad (f = 1)$$

Error variation S_e

$$S_e = S_T - S_\beta - S_{\beta \times N}$$
$$= 1{,}758{,}524.77 - 1{,}587{,}520.30 - 38{,}140.17 = 132{,}864.30 \qquad (f = 22)$$

Error variance V_e

$$V_e = S_e / 22 = 6{,}039.29$$
$$V_N = (S_T - S_\beta) / 23 = 7{,}434.98$$

S/N ratio η

$$\eta = 10 \log \frac{\frac{1}{r r_o}(S_\beta - V_e)}{V_N}$$
$$= 10 \log \frac{(1{,}587{,}520.30 - 6{,}039.29)}{14{,}080{,}263.32 \times 7{,}434.98} = -48.21 \text{ (dB)}$$

Sensitivity, S

$$S = 10 \log \frac{1}{r r_o}(S_\beta - V_e)$$
$$= 10 \log \frac{(1{,}587{,}520.30 - 6{,}039.29)}{14{,}080{,}263.32} = -9.50 \text{ (dB)}$$

Control factors B through H were assigned to columns 2 through 8, respectively. In each experimental run, the total amount was considered the standard amount, then was varied by +/– 30%. With two-level noise factors, six tests were conducted. In each test, temperature was measured four times for each time segment.

CALCULATION OF S/N RATIO AND SENSITIVITY

The target heat-generating time of the product is 24 hr. Since the amount of the ingredients is taken as one of the signal factors, the target heat-generating time will be different. They are:

M*1 : 24 x 0. 7 = 16.8 (hr)
M*2 : 24
M*3 : 24 x 1.3 = 31.2 (hr)

The average temperature was measured four times; at 1/4, 2/4, 3/4 and 4/4 of target heat-generating time. 30°C was subtracted from each temperature since the tests were made after setting the environmental temperature at 30°C by use of warm water as shown in Figures 1 and 2. Therefore, the amount of heat generation was calculated from the heat increase from 30°C .

To calculate the amount of heat generation in experiment No. 1 at M*1(16.4 hr) and M1(1/4), its heat-generating time is 16.4/4 = 4.2 hr. The average temperature during this 4.2 hr was measured, and 30°C was subtracted and it was multiplied by 4.2 hr to become Y1. Similarly, Y2, Y3 and Y4 were calculated from the average temperature during the time from the start to 8.4, 12.6 and 16.8 hr, respectively. In the cases of M*2 and M*3, target heat-generating time of 24 and 31.2 hr are used, respectively.

RESULTS

Table 6 shows the results of the calculations. Table 7 is the response table for the S/N ratio and sensitivity. Figures 3 and 4 show the effects of S/N ratio and sensitivity.

Table 6: S/N Ratio and Sensitivity

	e 1	B 2	C 3	D 4	E 5	F 6	G 7	H 8	S/N Ratio	Sensitivity
1	1	1	1	1	1	1	1	1	–48.21	–9.50
2	1	1	2	2	2	2	2	2	–51.98	–10.47
3	1	1	3	3	3	3	3	3	–53.65	–11.10
4	1	2	1	1	2	2	3	3	–53.38	–9.99
5	1	2	2	2	3	3	1	1	–52.12	–13.67
6	1	2	3	3	1	1	2	2	–52.04	–10.61
7	1	3	1	2	1	3	2	3	–55.76	–13.94
8	1	3	2	3	2	1	3	1	–51.42	–10.97
9	1	3	3	1	3	2	1	2	–50.92	–11.69
10	2	1	1	3	3	2	2	1	–50.96	–11.38
11	2	1	2	1	1	3	3	2	–50.91	–10.12
12	2	1	3	2	2	1	1	3	–47.81	–10.62
13	2	2	1	2	3	1	3	2	–52.14	–9.99
14	2	2	2	3	1	2	1	3	–52.58	–12.28
15	2	2	3	1	2	3	2	1	–50.82	–12.14
16	2	3	1	3	2	3	1	2	–52.92	–14.35
17	2	3	2	1	3	1	2	3	–50.28	–10.47
18	2	3	3	2	1	2	3	1	–52.81	–10.78

Table 7: Response Table for S/N Ratio

Factor	S/N Ratio			Sensitivity		
	Level 1	Level 2	Level 3	Level 1	Level 2	Level 3
B	–50.58	–52.18	–52.35	–10.53	–11.45	–12.03
C	–52.23	–51.55	–51.34	–11.53	–11.33	–11.16
D	–50.75	–52.10	–52.26	–10.65	–11.58	–11.78
E	–52.05	–51.39	–51.68	–11.20	–11.42	–11.39
F	–50.32	–52.10	–52.70	–10.36	–11.10	–12.55
G	–50.76	–51.97	–52.39	–12.02	–11.50	–10.49
H	–51.06	–51.82	–52.24	–11.41	–11.21	–11.40

Figure 3: S/N Ratio Response

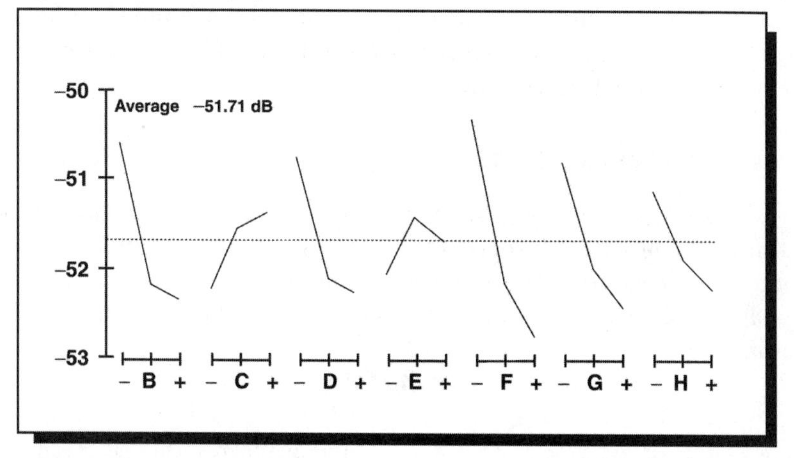

Figure 4: Sensitivity Response

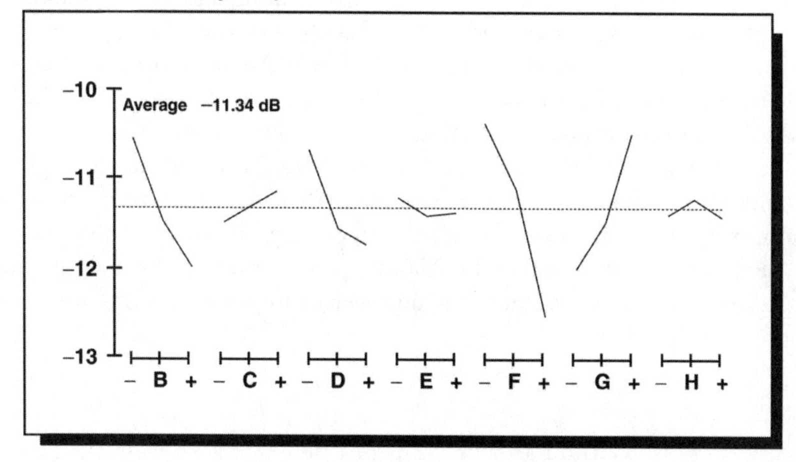

Confirmation

From the results of the S/N ratio, the optimum set of conditions were selected as B1-C3-D1-E2-F1-G1-H1. The results of the confirmation runs are shown in Table 8.

Table 8: Results of Confirmation

	S/N Ratio η (dB)			Sensitivity S (dB)		
	Optimum	Current	Gain	Optimum	Current	Gain
Estimation	–47.29	–53.23	5.94	–9.55	–11.62	2.07
Confirmation	–48.67	–52.58	3.91	–9.27	–11.83	2.56

The S/N ratio and sensitivity under the current and the optimum conditions were estimated using strong effects of B1, D1, F1 and G1. The confirmation shows fairly good reproducibility.

Results of Experimentation

It was considered ideal that the temperature generated is constant and the heat-maintaining time is proportional to the amount of ingredients. The target heat-maintaining time was divided into four segments. Letting the amount of heat generated in each segment be Y, the ideal function was evaluated by

$$Y = \beta MM^*$$

where M and M* are time and amount of ingredients, respectively. Generally, the amount of heat-generating ingredients and the resultant temperature are correlated. The lower the amount of heat-generating ingredients, the higher the initial peak temperature and the shorter the heat-maintaining time. On the other hand, the more that is put in, the lower the initial peak temperature and the longer the heat-maintaining time. Such trends were found in this study. Under the optimum condition, the air supply (permeability) was set at a lower level to lower the initial peak temperature. As a result, although the amount of heat-generating ingredients was 30% less than in the current condition, both showed almost the same temperature. It was probably because the sensitivity S was improved by 2.56 dB (1.803 times on the antilog scale). Therefore, the amount of heat per unit weight increased by 1.34 times (the square root of 1.803).

Comparing the heat-maintaining time on the same basis (M*2), the time under the optimum condition was a little shorter than in the current condition, resulting a smaller total amount of heat generation. But by adjusting the amount of ingredients, the heat-maintaining time can be set to the same length as the current product. The S/N ratio under the optimum condition was improved by 3.91 dB, becoming closer to the ideal function. Using the optimum condition as a base, it is planned to continue the study, and also to reduce the cost. The success of this study was beyond expectation, and the power of robust design approach was again realized.

This case study is contributed by Hiroshi Shimoda of Fumakilla, Ltd of Japan. Special thanks to Dr. Yano of The University of Electro-Communications, Mr. Ichigo of Hiroshima Kogyo Gijyutu Center and the members of Hiroshima District Quality Engineering Forum., Mr. Nakajima of Toa Gosei Co., Ltd. for their assistance and advice.

CHAPTER TEN

Energy Efficiency Compressors

GOLDSTAR CO., LTD. – SEOUL, KOREA

EXECUTIVE SUMMARY

BACKGROUND

International consensus was reached to take steps to reduce the emissions of both ozone layer depleters and "greenhouse" gases that are implicated in global warming. All chlorofluorocarbon (CFC) and HCFC compounds possess an ozone depletion potential (ODP) and global warming potential (GWP). There was a challenge to the refrigeration industry created by the need to reduce or even eliminate the emissions of all compounds with a significant ODP and/or GWP. Therefore, there is a need to develop higher efficiency compressors for the refrigeration industry in order to satisfy the growing need worldwide.

Robust design methodology was used to optimize the design process of valve parts of compressors that would affect the energy efficiency ratio (EER). Several parameters that influence refrigeration capacity were examined in order to determine the optimal design of valve parts of compressors.

PROCESS

The robust design method proved to be extremely helpful in reducing EER in optimizing valve design of compressors. Eight design parameters were tested under one noise factor, oil. Eighteen design configurations were tested where input energy in wattage was considered the signal and output energy in wattage was the response.

BENEFITS

- The dynamic S/N ratio approach proved to be extremely helpful in improving EER in optimizing valve design of the compressors.
- There is a 3.63-dB gain in S/N ratio, which is equivalent to reducing variance by half or more. With this improvement, efficiency has gone up by 3.3%. This is very significant in this industry.

- If this method is used widely to design valve parts of compressors, output capacity will be greatly increased by the refrigeration industry.
- The design is robust against variation of oil as noise.
- Moreover, this design optimization satisfies needs of the society to control environmental pollution.
- A series of studies resulted in compressor designs that were regarded as grade A internationally. Before these developments, compressors in the Korean industry were regarded as grade C.

CASE STUDY

INTRODUCTION

The past 10 years have seen an international consensus take steps to reduce emissions of both ozone layer depleters and "greenhouse" gases that are implicated in global warming. All these CFC and HCFC compounds possess an ozone depletion potential (ODP) and a global warming potential (GWP).

This paper is concerned with the challenge presented to the refrigeration industry by the need to reduce or even eliminate the emission of all compounds with a significant ODP and/or GWP. There is a great need to develop a more efficient compressor for the refrigeration industry in order to satisfy the growing worldwide demand. The study presented in this paper utilized the Taguchi design of experiment method to optimize the design process of valve parts of compressors that would affect the energy efficiency ratio (EER). Several parameters (or factors) that influence refrigeration capacity were examined in order to determine the optimization of the design process of valve parts. The signal-to-noise (S/N) ratio is used as a performance index for each experimental combination to analyze the data in order to determine the significant factor and level at which the improvement could be achieved when design parameters were changed or adjusted.

THE ENGINEERED SYSTEM

Ideal Function

The most effective ideal function for the compressor system is defined as

Input Signal (M) = Input Energy (W)
Out Response (y) = Output Energy (W)

The above definition of the ideal function is considered by the engineers to be the most appropriate interpretation of the system input/output energy transfer. The direct measurement of the planetary input/output energy by the experimental apparatus in this test is shown in Figure 1 and the various types of valve parts (including suction and discharge valve, head seat) are assembled into the whole system of the compressor.

Control and Noise Factors

There are a number of factors as illustrated in Figure 2 that can affect the EER. In Table 1, the most important eight factors based on engineers' experience have been chosen as control factors for the experiment design. The inner array contains all control factors, i.e., the design parameters, whose levels can be set by the design. The L_{18} orthogonal array arrangement chosen in this study is shown in Table 2.

For the outer orthogonal array, the voltages are used as the signal factor and N1, N2 are used as the compounded noise levels, which were chosen to anticipate the effect of manufacturing, variability and application conditions.

Figure 1: Experimental Apparatus and Value-Related Parts

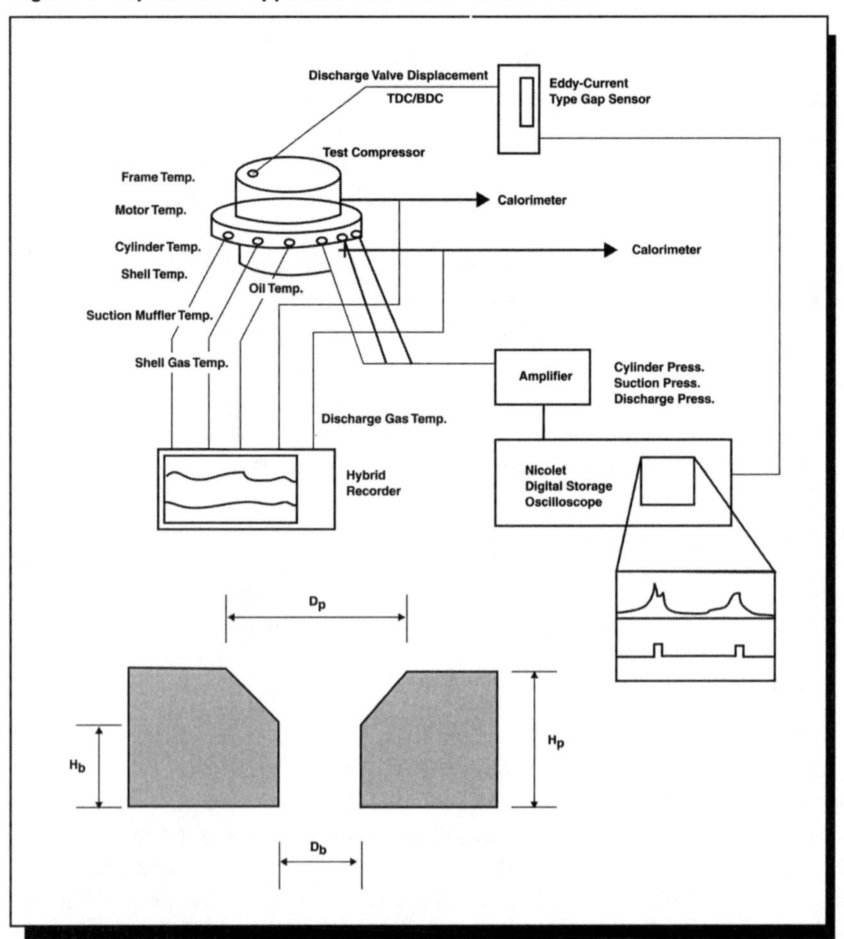

Figure 2: Response Graph

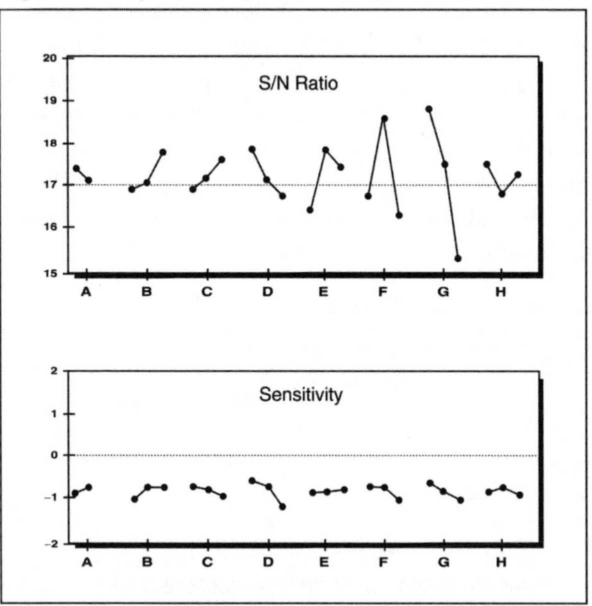

Table 1: Control and Noise Factors

Control Factors

Factors	Level 1	Level 2	Level 3
1. Height 1	2.0	4.0	---
2. Diameter 1	4.0	5.0	6.0
3. Port Dia. 1/Bore Dia. 1	1.5	2.0	2.5
4. Diameter 2	3.0	5.0	7.0
5. Port Dia. 2/Bore Dia. 2	1.5	1.8	2.1
6. Thickness 1	0.15	0.20	0.25
7. Thickness 2	0.20	0.25	0.30
8. Height 2	2.00	2.05	2.10

Signal and Noise Factors

Factors	Level 1	Level 2	Level 3
1. Signal Factor (RPM)	210 V	220 V	230 V
2. Noise Factor (oil)	Little	Much	---

EXPERIMENT AND MEASUREMENT

The experimental apparatus used in this test is shown in Figure 1. Various types of valve parts are assembled into the test compressor and the input/output electric energy (EER) as measured by the calorimeter. We also measured the temperature and pressure of many parts and we recorded the data in a recorder and an oscilloscope. Also we discharged valve displacement by using an eddy-current type of gap sensor to measure cylinder volume. Table 2 shows the experimental data.

Table 2: Experimental Layout and Data

Factor Level	A	B	C	D	E	F	G	H
1	2.0	4.0	1.5	3.0	1.5	0.15	0.20	2.00
2	4.0	5.0	2.0	5.0	1.8	0.20	0.25	2.05
3	---	6.0	2.5	7.0	2.1	0.25	0.30	2.10

	Input						Output					
	Noise 1			Noise 2			Noise 1			Noise 2		
	M_1	M_2	M_3	M_1	M_2	M_3	M_1	M_2	M_3	M_1	M_2	M_3
1	9.84	9.93	10.06	9.91	10.00	10.12	9.53	9.53	9.52	9.41	9.40	9.45
2	9.24	9.36	9.50	9.32	9.41	9.55	8.54	8.55	8.58	8.67	8.60	8.57
3	8.50	8.65	8.81	8.68	8.81	8.95	6.81	6.91	6.84	7.23	7.22	7.20
4	9.84	9.94	10.07	9.85	9.94	10.06	9.39	9.39	9.39	9.44	9.47	9.38
5	9.63	9.73	9.86	9.67	9.78	9.91	9.23	9.24	9.27	9.34	9.36	9.31
6	8.92	9.06	9.21	9.05	9.09	9.28	7.91	8.04	7.99	8.21	8.23	8.22
7	9.11	9.22	9.35	9.20	9.31	9.46	8.26	8.22	8.19	8.47	8.45	8.45
8	8.93	9.01	9.19	9.01	9.09	9.27	7.92	7.92	7.94	7.86	8.14	7.91
9	9.93	10.03	10.15	10.09	10.18	10.29	9.67	9.73	9.70	9.77	9.83	9.83
10	9.10	9.21	9.36	9.16	9.26	9.40	8.13	8.14	8.12	8.19	8.22	8.24
11	9.34	9.46	9.07	9.57	9.67	9.79	8.43	8.39	8.43	8.58	8.60	8.56
12	9.36	9.53	9.66	9.45	9.55	9.68	8.85	8.88	8.89	8.96	8.97	9.02
13	9.51	9.67	9.75	9.54	9.70	9.75	8.81	9.39	8.93	8.99	9.39	9.27
14	9.04	9.15	9.28	9.08	9.20	9.34	8.23	8.25	8.25	8.39	8.40	8.37
15	9.50	9.61	9.75	9.59	9.69	9.82	8.82	8.80	8.80	8.65	8.76	8.74
16	9.11	9.25	9.38	9.17	9.27	9.41	8.48	8.48	8.46	8.51	8.43	8.48
17	10.08	10.15	10.27	10.12	10.20	10.31	9.93	9.78	9.80	9.89	9.90	9.90
18	9.29	9.41	9.55	9.36	9.47	9.61	8.59	8.66	8.70	8.87	8.87	8.81
OP	9.90	9.99	10.11	9.92	10.01	10.13	9.78	9.80	9.78	9.83	9.83	9.86

Results and Data Analysis

The dynamic analysis of the S/N ratio is based on the following equation:

$$\eta = 10 \log \frac{\frac{1}{2r}(S_\beta - V_e)}{V_N}$$

Response tables for the S/N ratio and sensitivity are shown in Tables 3 and 4, respectively. Figure 2 shows the response graph. After some trade-off between S/N ratio and sensitivity in the process of the two-step optimization, the optimal design was selected as $A_2B_3C_1D_1E_3F_1G_1H_2$.

Table 3: Response Table for S/N Ratio

Level	A	B	C	D	E	F	G	H
1	17.41	16.94	16.93	17.87	16.39	16.77	18.84	17.56
2	17.07	17.01	17.19	17.11	17.84	18.65	17.54	16.84
3	---	17.78	17.61	16.76	17.50	16.31	15.35	17.33

Table 4: Response Table for Sensitivity

Level	A	B	C	D	E	F	G	H
1	–0.8269	–0.9535	–0.7001	–0.5559	–0.8128	–0.6602	–0.5684	–0.7720
2	–0.7254	–0.6878	–0.7448	–0.6445	–0.7725	–0.6976	–0.8204	–0.7221
3	---	–0.6872	–0.8836	-1.1280	–0.7432	–0.9707	–0.9396	–0.8344

Confirmation

The S/N ratio and sensitivity for the current design and the optimal design were predicted. Table 5 shows the prediction and actual results from the confirmation run.

The S/N ratio was improved by 3.36 dB and sensitivity by 0.278 dB. This translates to a reduction of variance by more than half and an increase in β by 3.3%. An increase in β by 3.3% is very significant in this industry.

Table 5: Prediction and Confirmation

Condition	S/N Ratio		Sensitivity	
	Prediction	Confirmation	Prediction	Confirmation
Current	17.85	17.16	–0.3333	–0.4516
Optimum	19.80	20.79	–0.067	–0.1737
Gain	1.95	3.63	0.2663	0.2779

Conclusion

The dynamic analysis of S/N ratio is extremely useful to improve the energy efficiency ratio. The optimal condition can improve the EER greatly during the developing process. If this skill is used widely to design valve parts of compressors, output capacity will be greatly increased in the refrigeration industry.

This case study is contributed by Sung Keun Park, Soo Youn Ji, Seog Jong Hong, Keum Sik Im, Moo Soo Noh of Goldstar Co., Ltd., – Seoul, Korea. The authors would like to thank Ki Chul Choi for his helpful experimental support.

CHAPTER ELEVEN

Optimization of a Cosite Mitigation Receiver

ITT AEROSPACE/COMMUNICATIONS DIVISION – USA

EXECUTIVE SUMMARY

BACKGROUND

Wireless communications are vital to everyday civilian and military life. Voice and data transmitted over any distance must be received with little to no error. In order to attain the range of communications necessary, communications systems are designed to utilize high power transmitters and sensitive receivers along with other RF processing techniques.

In many situations several radio transmitter/receiver combinations must be co-located to service the communications needs. An instance of this is found at installations such as at a military command post or vehicular installation which must oversee the actions of hundreds or even thousands of troops. These troops are generally widely dispersed and organized to communicate on different channels at the same time, requiring the commander to operate with several radios simultaneously. This presents a serious challenge to radio design engineers since in this configuration it can never be fully determined how many transmitters will be required to be active at the same time because any number of receivers must be tuned to distant stations. Cosite interference is the term used to describe the situation where a high power transmitter is active and interfering with a co-located receiver.

There are several methods used to mitigate cosite interference in communication systems design. One easy method is simple transmitter/receiver separation. This, however, is of limited use to the command post or vehicle that must be organized in a small area. Other methods include filtering, active cancellation and digital signal processing techniques or combinations of these.

PROCESS

This paper presents the optimization of the receiver section of a cosite mitigation system which utilizes a combination of active and passive filtering and active cancellation. The design engineers recognized that there were many combinations of the nominal values of the critical parameters of the filters, gain stages and cancellation that were possible to obtain the nominal performance of the system. They also recognized that the environment, an airborne command and control post, was outfitted with multiple transceivers of several types. In addition, the combination of communication range, channel frequencies and the number of active co-located transmitters was not controllable by the designer and the operator. This presented an excellent opportunity to perform parameter design utilizing Taguchi Methods™.

Parameter diagram analysis of the system was used to capture the main function of the system and identify the control and noise factors. Essentially, if functioning ideally, the system would only pass the desired received signal no matter how many other interfering signals were in the environment. Further analysis showed that two elements of the ideal system function needed to be measured to ensure proper optimization. The first function was of the dynamic type with the ratio of desired-to-undesired signal at the system input (input carrier to noise or C/N) as the abscissa and the ratio of the desired-to-undesired signal at the output (output C/N) as the ordinate. The zero-point-proportional dynamic SNR was used to analyze the factor effect of this ideal function. The second function resulted from the observation that there is a certain threshold beyond which degradation of the received signal by cosite interferences increases at an extremely rapid rate to the point where the channel is essentially blocked. Thus to design a system that ideally resulted in no blocks was desired. An omega transform of percent blocks and the smaller-the-better SNR were used to analyze this ideal function.

Combinations of five control factor values were setup using an L_{18} with columns 1, 7 and 8 empty. The values of six noise factors were combined using an L_{18} with columns 1 and 8 empty. The system was simulated utilizing a mathematical model and the control and noise value settings were programmed as input variables into the simulation. In this way the entire simulation could be run hands-off. Output data was collected and analyzed separately.

Following the analysis of factor effects for both ideal functions, optimum parameter values were set and a confirmation run was performed. Confirmation results were very good and the final parameter value selections were made. These values were, in several cases, those chosen as initial settings proving the initial integrity of the design. One of the most interesting results was that the input noise figure was a weak contributor to performance. It was believed by the customer to be one of the most critical parameters.

Benefits

- First time success at flight test saving thousands of man-hours of potential redesign and retest by both the contractor design/test team and the customer.
- Determination that system integration could begin with low risk saving hundreds of man-hours of "standard" analysis and/or test/analyze/fix time.
- Noise factor requirements could be relaxed resulting in cost savings for the input low noise amplifier.
- Developed baseline simulation/optimization technology for this type of system design.

CASE STUDY

Introduction

ITT Aerospace/Communications Division develops and produces two major product lines, tactical communications equipment and satellite borne navigation equipment and atmospheric sensors. The tactical communications business primarily targets at the needs of the U.S. Army. Systems currently in development and production include ground, airborne and export versions of SINCGARS, a secure VHF voice and data radio with antijam capability, high capacity data radios to support the Army's effort to digitize the battlefield with a tactical Internet structure and a new product line aimed at mitigating the effects of multiple VHF/UHF radios co-located with each other. This parameter design effort was undertaken to address design issues related to the first generation of cosite equipment being developed at ITT A/CD known as the Advance Interference Mitigation System (AIMS).

Many tactical communications devices are configured within a command post environment when fielded. This results in several transmitters and receivers installed in one location in close proximity to one another. When two-way radios, of any type, are co-located with each other, the high output power of a local transmitter tends to interfere with the highly sensitive local receiver (Figure 1). This is due to the low levels of antenna-to-antenna isolation available in a typical installation. Frequency management can provide enough isolation to protect single-channel radio applications, but is ineffective when frequency-hopping radios are used. The ITT cosite mitigation system is targeted at reducing interference effects for co-located frequency-hopping radio installations.

Cosite mitigation systems attempt to permit communications of all co-located radios as if they are in a noncosited situation. Previous cosite mitigation systems attempted to use a single mitigation method to eliminate the inter-

Figure 1: Receiver Situation Co-Located with Other Transmitters

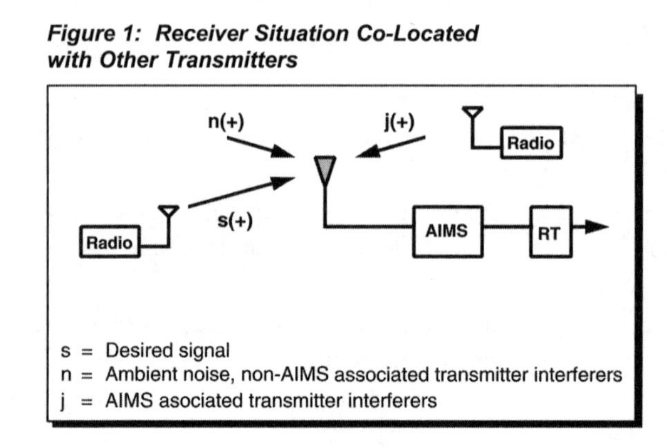

fering signals. These methods were based on either filtering or active cancellation. Filtering removes interference through the use of narrow bandpass filters that are tuned into either the transmitter or receiver in an attempt to isolate these signals from each other. In a system that must operate with frequency-hopping radios, these filters must be able to switch center frequency in reaction to the tune sequence generated by the host radio. Active cancellation utilizes phase shift techniques to cancel the interfering signal in the receive path. Individually, each of these techniques is effective to a degree. ITT A/CD has defined a system which combines these techniques into a single system. The combination of techniques results in a more effective cosite system.

Background

Cosite mitigation systems can provide both suppression of undesired energy emanating from local transmitters and signal conditioning (enhancement) of the desired signal. From the frame of reference of a local receiver any signal from a radio that is attempting to communicate with it is desired and any other signals are undesired. We therefore classify, from the receiver's perspective, two classes of signals: desired and undesired. The portion of the cosite system allocated to operating on the transmitted signal is called the cosite transmitter and that portion dedicated to operating on the received signal the cosite receiver. This effort focused on the cosite receiver portion of the system.

The cosite receiver is comprised of an active canceller, a receive filter, a low noise amplifier (LNA) and an attenuator. A block diagram of the receiver is shown in Figure 2.

Prior to the initiation of this parameter design effort, standard design methodologies were being employed to determine the best values for the performance of subsystems. A mathematical model was developed that sim-

Figure 2: Cosite Receiver Block Diagram

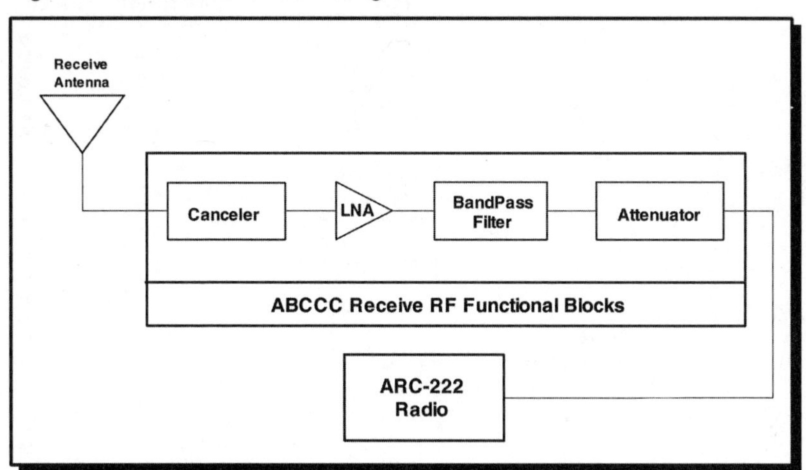

ulated the system and its operating environment. The model includes the effects of a desired signal path loss using the Longley-Rice propagation model, antenna-to-antenna isolation and environmental noise. It was used to select the parameter values for each of the major components of the system and to evaluate the effectiveness of the AIMS architecture. With this approach, performance levels and their associated effectiveness were determined for some of the subsystem parameters. However, other critical subsystem parameters had not been evaluated and it was somewhat unclear how best to proceed to determine the values and achieve optimal performance within the bounds of the technology available. Intuition also indicated that some of the system's level requirements were overspecified and could be relaxed without detriment to the overall effectiveness of the system when installed.

Objectives

The objective of the AIMS system is to meld technologies in order to achieve greatly improved interference mitigation effectiveness. The parameter design effort was primarily designed to select the parameters for the AIMS front end LNA to optimize the performance of the receiver. It also acted to confirm that the selection of values for cancellation and filters is optimum for overall cosite mitigation performance.

As stated previously, cosite mitigation systems attempt to permit communications of all co-located radios as if they were in a noncosited situation. The main function of this system is therefore the amplification of the signals that are desired and the attenuation of the signals that are undesired. There exists only one desired signal at a time for any given receiver. The undesired signals, however, consist of up to three interfering transmitters, interference from

Figure 3: Parameter Diagram

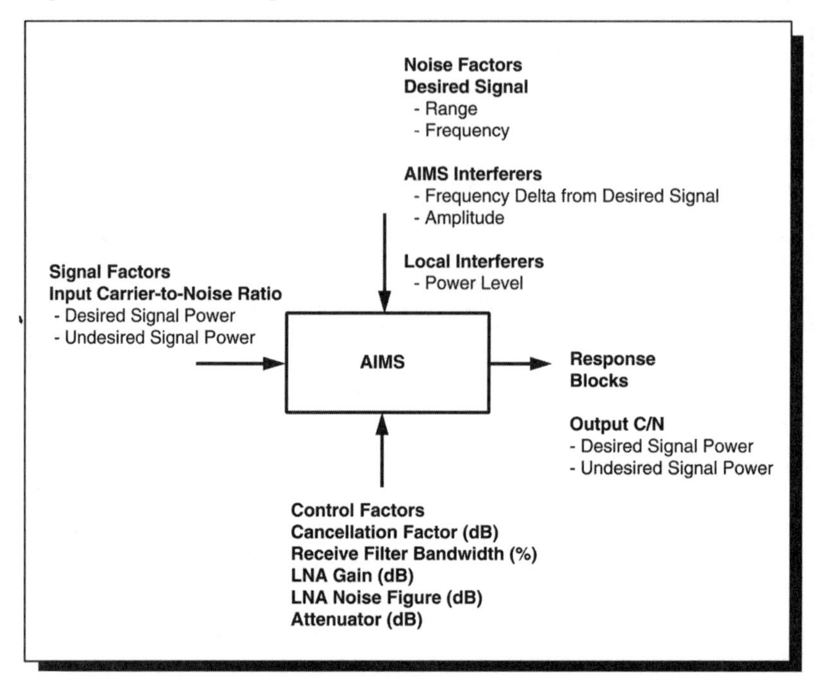

other communications systems and environmental/thermal noise. The parameter diagram for the AIMS system is shown in Figure 3. A description of each of the parameters is in Table 3.

Noise Factors

Range: The received signal is the desired signal of interest. The range of the desired signal transmitter from the AIMS receiver is the starting point for the calculation of the received signal amplitude. Power at the receive antenna is calculated based upon a Longley/Rice propagation model. A typical transmitting radio is used as a source, which for this design effort is a manpack.

Desired Signal Frequency: This is the RF frequency of the desired signal in megahertz. These values range over the SINCGARS operating band. A range of frequencies was chosen to ensure optimization over the entire band.

Interferer Frequency: This is the frequency of the interfering signals in megahertz. They are defined as an offset from the frequency of the desired signal.

Interferer Output Power: This is the power level of the interferer. The reference point for this level is at the base of the interferer transmitting antenna. Path loss to the receiving antenna is then accounted for based on measured antenna-to-antenna isolation.

Local Interferer Power Level: This is the sum of the broadband and environmental noise at the aircraft antenna.

Control Factors

Cancellation: The amount (in dB) of attenuation provided against the received interfering signal. Three values were chosen to span the available technology. These values are applied to all of the active interferers.

Receive Filter Bandwidth: The filter BW (in percent) of the receive filter at its 3-dB points. These filters will reduce the noise power from the AIMS interferers as well as other on and off platform interferers. Only the known interferers are accounted for. A fixed bandwidth x insertion loss product model is used therefore as the filter BW is decreased the insertion loss increases.

Front End LNA: The LNA at the front end of the system provides the dominant effect in establishing the noise figure (NF) of the system, and hence directly influences the overall system sensitivity. Three gains and NFs for candidate LNAs will be utilized.

Attenuator: A fixed attenuator is used for VSWR matching and to establish power levels, thereby protecting the PA and the radio. A three-position switch controls the selection of either a 0-, 4- or 10-dB pad.

Response Variables

C/N: The carrier-to-noise ratio is calculated by calculating the ratio of the power level of the desired received signal with the sum of the power of the interfering signal, local noise, and receiving noise

Blockage: When the frequency of an interfering signal falls within the bandwidth of the receiver channel, no mitigation is possible and the channel is considered "blocked." This metric is calculated by summing the number of blocks that occur during an experiment scenario.

The effects of the control parameters on system performance were studied using an ideal function that plotted SQRT input carrier/noise vs. SQRT output carrier/noise (see Figure 4). SQRT C/N was used due to its being better adapted to the analysis. The efforts focused on linearizing this plot and maximizing the slope. The slope and variance from the slope were calculated during the analysis from a regression. An S/N ratio was then calculated from the slope and variance. A metric that counted the number of blocks was also used. This metric was minimized.

Figure 4: Ideal Function or C/N Response

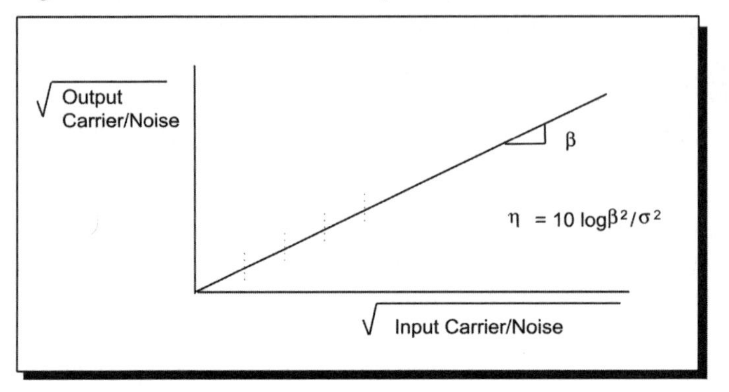

Experiment Layout

Levels for the noise factors are shown in Table 1. The levels for control factors are in Table 2. An L_{18} orthogonal array was constructed from the noise factors. The desired signal range was treated as an outside factor as a convenience to the simulation. Thus the noise orthogonal array was run three times for each row corresponding to each range. The noise factor OA is shown in Table 3. The L_{18} array constructed from the control factors is shown in Table 4.

Table 1: Noise Factor Levels

Factor	Level 1	Level 2	Level 3
Desired Signal Range (km)	45	80	115
Desired Signal Frequency (MHz)	40	60	75
Interferer #1 Offset Frequency (MHz)	0.55	0.35	0.40
Interferer #2 Offset Frequency (MHz)	–1.10	–0.70	–0.80
Interferer #3 Offset Frequency (MHz)	1.65	1.05	1.20
Interferer #1 & 2 Amplitude (watts)	12.5	25	40
Interferer #3 Amplitude (watts)	12.5	25	40

Table 2: Control Factor Levels

Factor	Level 1	Level 2	Level 3
Cancellation (dB)	Low	Mid	High
Rev Filter BW (%)	Low	Nominal	Wide
LNA Gain (dB)	Low	Nominal	Wide
LNA Noise Figure (dB)	Low	Nominal	High
Attenuator (dB)	Low	Nominal	High

Table 3: Noise Array Test Log (L_{18})

Test No.	Dupl. of Test No.	A: Col. 2 Freq. ds, f (d) MHz	B: Col. 3 Radio 1 Freq. MHz	C: Col. 4 Radio 2 Freq. MHz	D: Col. 5 Radio 3 Freq. MHz	E: Col. 6 Ampl. Radio 1 & 2 dB	F: Col. 7 Ampl. Radio 3 dB	Col. 1 Empty
1		1:40.00	1:40.35	1:40.55	1:40.40	1:12.50	1:12.50	
2		1:40.00	2:38.95	2:38.35	2:38.80	2:25.00	2:25.00	
3		1:40.00	3:43.15	3:44.95	3:43.60	3:40.00	3:40.00	
4		2:60.00	1:60.35	1:60.55	2:58.80	2:25.00	3:40.00	
5		2:60.00	2:58.95	2:58.35	3:63.60	3:40.00	1:12.50	
6		2:60.00	3:63.15	3:64.95	1:60.40	1:12.50	2:25.00	
7		3:75.00	1:75.35	2:73.35	1:75.40	3:40.00	2:25.00	
8		3:75.00	2:73.95	3:79.95	2:73.80	1:12.50	3:40.00	
9		3:75.00	3:78.15	1:75.55	3:78.60	2:25.00	1:12.50	
10		1:40.00	1:40.35	3:44.95	3:43.60	2:25.00	2:25.00	
11		1:40.00	2:38.95	1:40.55	1:40.40	3:40.00	3:40.00	
12		1:40.00	3:43.15	2:38.35	2:38.80	1:12.50	1:12.50	
13		2:60.00	1:60.35	2:58.35	3:63.60	1:12.50	3:40.00	
14		2:60.00	2:58.95	3:64.95	1:60.40	2:25.00	1:12.50	
15		2:60.00	3:63.15	1:60.55	2:58.80	3:40.00	2:25.00	
16		3:75.00	1:75.35	3:79.95	2:73.80	3:40.00	1:12.50	
17		3:75.00	2:73.95	1:75.55	3:78.60	1:12.50	2:25.00	
18		3:75.00	3:78.15	2:73.35	1:75.40	2:25.00	3:40.00	

Table 4: ABCCC Orthogonal Array Test Log (L_{18})

Test No.	Dupl. of Test No.	A: Col. 1 Radio Noise	B: Col. 2 Cancellation (dB)	C: Col. 3 RCV Filter BW (%)	D: Col. 4 LNA NF	E: Col. 5 LNA Gain	F: Col. 6 Attenuator	Col. 7 Empty	Col. 8 Empty
1		1:ARC-222	1:low	1:narrow	1:low	1:low	1:low		
2		1:ARC-222	1:low	2:nominal	2:nominal	2:nominal	2:nominal		
3		1:ARC-222	1:low	3:wide	3:high	3:high	3:high		
4		1:ARC-222	2:mid	1:narrow	1:low	2:nominal	2:nominal		
5		1:ARC-222	2:mid	2:nominal	2:nominal	3:high	3:high		
6		1:ARC-222	2:mid	3:wide	3:high	1:low	1:low		
7		1:ARC-222	3:high	1:narrow	2:nominal	1:low	3:high		
8		1:ARC-222	3:high	2:nominal	3:high	2:nominal	1:low		
9		1:ARC-222	3:high	3:wide	1:low	3:high	2:nominal		
10		1:ARC-222	1:low	1:narrow	3:high	3:high	2:nominal		
11		1:ARC-222	1:low	2:nominal	1:low	1:low	3:high		
12		1:ARC-222	1:low	3:wide	2:nominal	2:nominal	1:low		
13		1:ARC-222	2:mid	1:narrow	2:nominal	3:high	1:low		
14		1:ARC-222	2:mid	2:nominal	3:high	1:low	2:nominal		
15		1:ARC-222	2:mid	3:wide	1:low	2:nominal	3:high		
16		1:ARC-222	3:high	1:narrow	3:high	2:nominal	3:high		
17		1:ARC-222	3:high	2:nominal	1:low	3:high	1:low		
18		1:ARC-222	3:high	3:wide	2:nominal	1:low	2:nominal		

Experiment Results and Analysis

The results of each experiment were gathered into a spreadsheet from the simulation. A regression was run for each of the SQRT input C/N vs. SQRT output C/N data groups. From this slope (B) and the variance from the slope (s 2) were obtained. An example of this data and the regression line are shown in Figure 5.

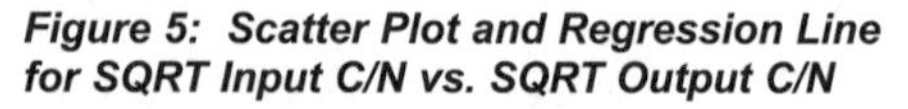

Figure 5: Scatter Plot and Regression Line for SQRT Input C/N vs. SQRT Output C/N

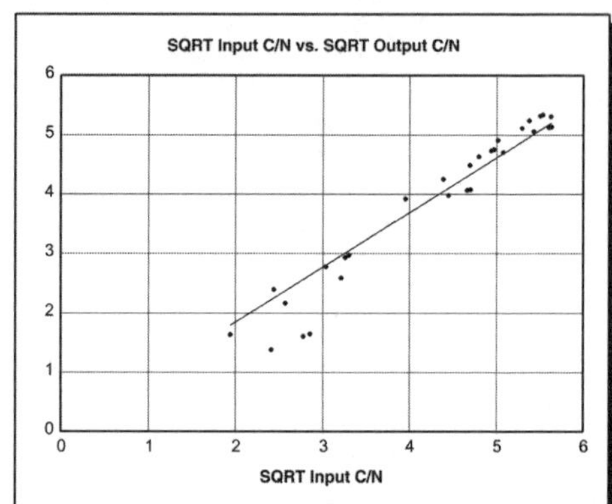

The B and s 2 data were used to calculate S/N for each SQRT C/N step in the control matrix. Table 5 shows the S/N for SQRT C/N for each experiment. The S/N for blocks was calculated by the following equation:

$$\eta = 10 \log[(1-p) / p]$$

where
p = percentage of blocks

The factor effects input data in Table 5 were further analyzed to separate out the factor effects for each control. Summaries of the factor effects for both SQRT C/N and blocks are shown in Table 6 and plotted against the control factor in Figures 6 and 7.

Table 5: ABCCC Control Orthogonal Array Factor Effects Input

Expt. No.	C/N S/N Ratio dB	C/N Sensitivity dB	SQRT C/N S/N Ratio dB	SQRT C/N Sensitivity dB	No. of Blocks S/N Ratio dB	No. of Blocks Sensitivity Count
1	-52.6	-11.8	-14.8	-5.1	1.0	24.0
2	-49.4	-8.2	-11.7	-3.5	-1.9	33.0
3	-48.4	-7.2	-10.7	-3.1	-4.1	39.0
4	-52.2	-11.3	-14.4	-4.9	8.9	6.0
5	-49.3	-8.0	-11.6	-3.4	4.1	15.0
6	-49.1	-7.9	-11.5	-3.4	0.0	27.0
7	-58.1	-23.5	-20.0	-10.3	23.4	0.0
8	-48.3	-7.2	-10.7	-3.1	8.9	6.0
9	-44.1	-4.3	-7.5	-1.9	3.0	18.0
10	-48.8	-7.6	-11.2	-3.2	-1.0	30.0
11	-55.8	-17.5	-17.8	-7.6	4.1	15.0
12	-46.3	-5.6	-9.3	-2.4	-4.1	39.0
13	-46.2	-5.6	-8.9	-2.4	1.9	21.0
14	-52.3	-11.4	-14.5	-4.9	6.9	9.0
15	-51.6	-10.5	-13.8	-4.5	6.9	9.0
16	-56.3	-18.7	-18.4	-8.1	23.4	0.0
17	-43.9	-4.2	-7.6	-1.8	1.9	21.0
18	-50.3	-9.0	-12.6	-3.9	23.4	0.0
Avg.	-50.2	-10.0	-12.6	-4.3	5.9	17.0
Max.	-43.9	-4.2	-7.5	-1.8	23.4	39.0
Min.	-58.1	-23.5	-20.0	-10.3	-4.1	0.0
Range	14.1	19.3	12.5	8.5	27.5	39.0

The factor effects analysis was used to choose the best combination of factor settings for the confirmation experiments and the design. This involved a session with the system experts to study this data and understand the manner in which certain items interacted with system performance in the context of the experiment design and simulation model. Trade-offs were made between the performance of the system for effectiveness against interferers for C/N and blockage. Blocks were determined to be the dominant cause of communications errors and were thus given a higher weight. Table 7 summarizes the factor settings for the baseline and two confirmation trials.

Table 6: ABCCC Control Orthogonal Array Factor Effects Table

Factor Name	Factor Levels	C/N S/N Ratio dB	C/N S/N Ratio F	C/N Sensitivity dB	C/N Sensitivity F	SQRT C/N S/N Ratio dB	SQRT C/N S/N Ratio F	SQRT C/N Sensitivity dB	SQRT C/N Sensitivity F	No. of Blocks S/N Ratio dB	No. of Blocks S/N Ratio F	No. of Blocks Sensitivity dB	No. of Blocks Sensitivity F
A: Radio Noise	A1: ARC-222	-50.16	0.0	-9.94	0.0	-12.55	0.7	-4.28	0.0	4.79	3.2	18.67	1.0
	A2: ARC-222	-50.17		-10.02		-12.67		-4.32		7.05		16.00	
B: Cancellation	B1: low	-50.21	0.6	-9.65	3.0	-12.59	1.9	-4.15	3.0	-1.01	47.9	30.00	24.1
	B2:: med	-50.12		-9.13		-12.46		-3.91		4.79		14.50	
	B3: high	-50.16		-11.17		-12.79		-4.85		13.98		7.50	
C: RCV Filter BW (%)	C1: low	-52.37	1280.0	-13.10	22.3	-14.61	235.9	-5.67	20.0	9.59	8.5	13.50	3.4
	C2: nominal	-49.83		-9.41		-12.32		-4.05		4.00		16.50	
	C3: high	-48.30		-7.44		-10.90		-3.19		4.17		22.00	
D: LNA NF	D1: low	-50.03	32.9	-9.95	0.0	-12.67	3.5	-4.29	0.0	4.29	2.6	15.50	0.5
	D2: nominal	-49.93		-10.00		-12.36		-4.32		7.79		18.00	
	D3: high	-50.54		-9.99		-12.81		-4.29		5.68		18.50	
E: LNA Gain	E1: low	-53.02	3003.6	-13.53	36.7	-15.21	543.6	-5.85	32.3	9.78	17.0	12.50	6.5
	E2: nominal	-50.69		-10.25		-13.06		-4.41		6.99		15.50	
	E3: high	-46.78		-6.17		-9.57		-2.64		0.98		24.00	
F:Attenuator	F1: 0	-47.75	2357.0	-7.07	38.3	-10.45	430.6	-3.04	34.1	1.60	13.7	23.00	4.8
	F2: 4	-49.51		-8.64		-11.99		-3.70		6.54		16.00	
	F3: 10	-53.23		-14.24		-15.40		-6.17		9.62		13.00	
Average		**-50.17**		**-9.98**		**-12.61**		**-4.30**		**5.92**		**17.33**	
Pooled Err Var		**0.02**		**2.22**		**0.09**		**0.48**		**7.15**		**33.00**	

Figure 6: Factor Effects for SQRT C/N

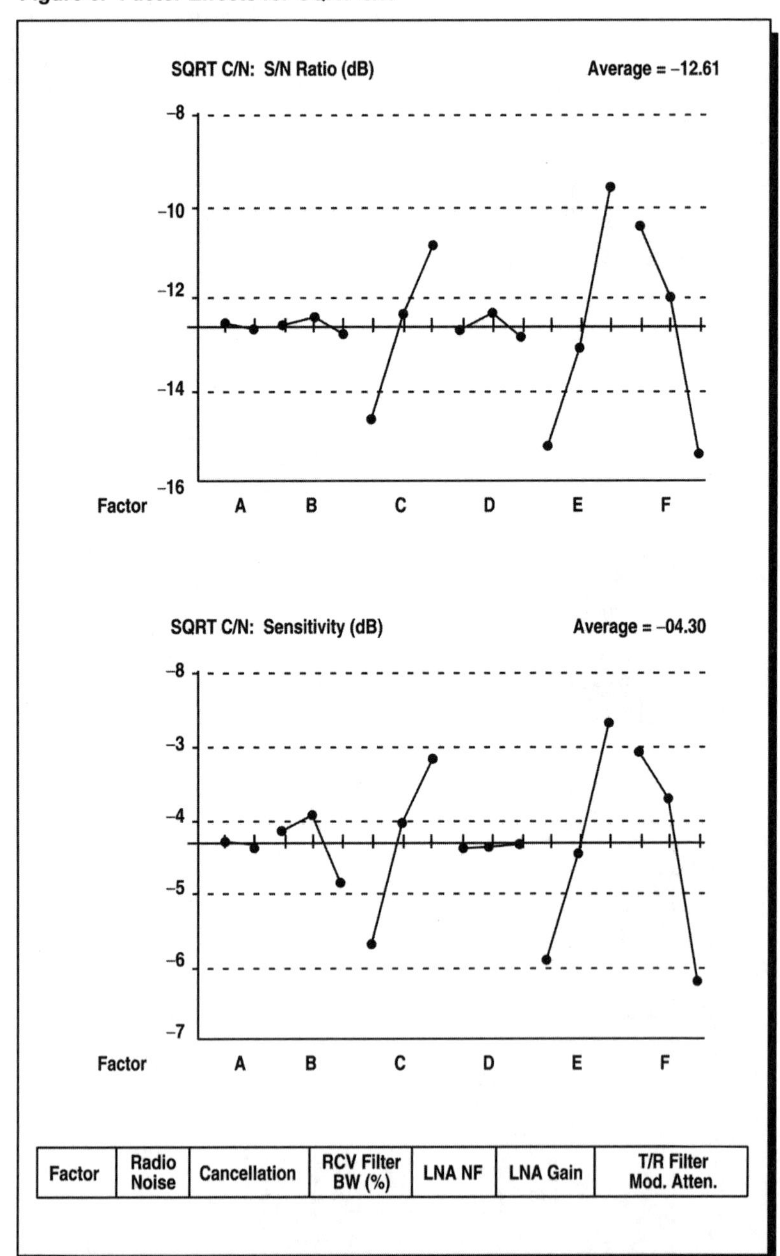

Figure 7: Factor Effects for Blocks

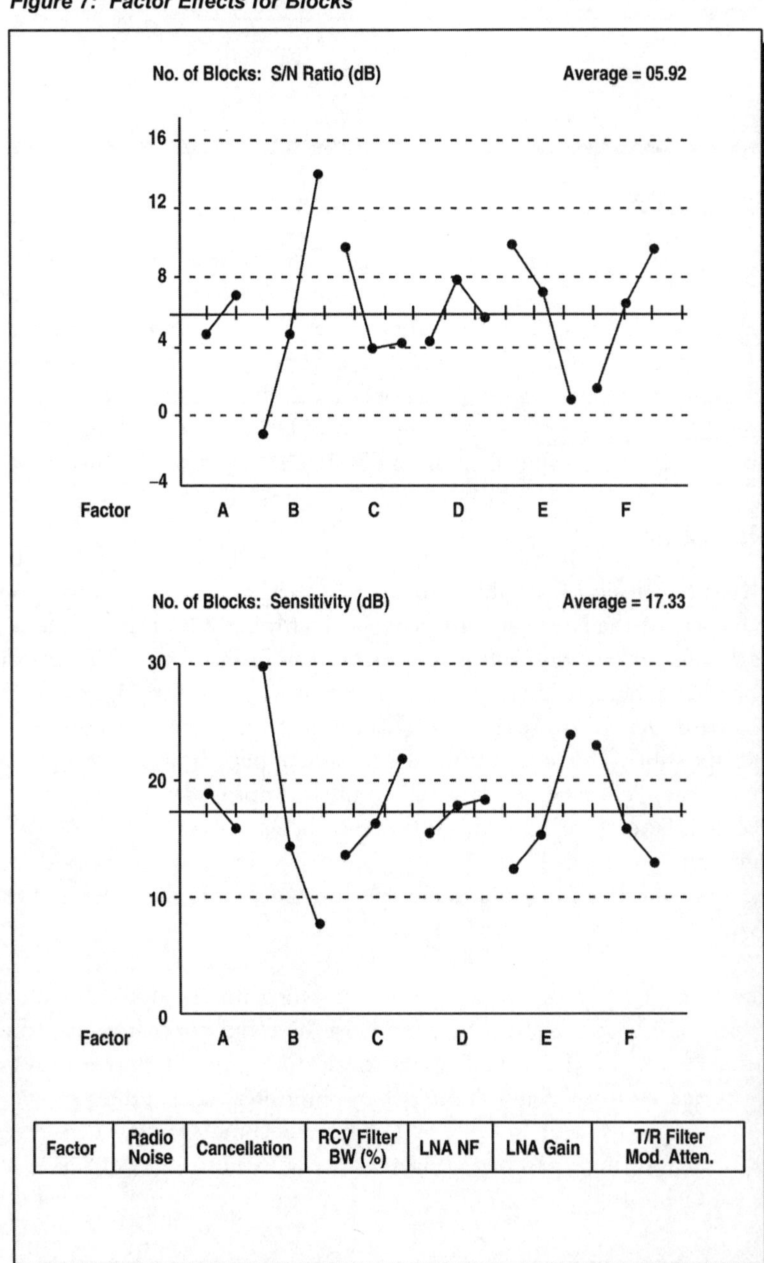

Factor	Radio Noise	Cancellation	RCV Filter BW (%)	LNA NF	LNA Gain	T/R Filter Mod. Atten.

Table 7: ABCCC Control Orthogonal Array Prediction Summary

Combination Name	Factor Settings	Intret Include	C/N S/N Ratio dB	C/N Sensitivity dB	SQRT C/N S/N Ratio dB	SQRT C/N Sensitivity dB	No. of Blocks S/N Ratio dB	No. of Blocks Sensitivity dB
Baseline	A1, B3, C1, D2, E2, F1	No	-50.2	-11.6	-12.8	-5.1	15.1	9.5
Trial 1	A1, B3, C1, D2, E2, F2	No	-52.0	-13.2	-14.3	-5.7	20.1	2.5
Trial 2	A1, B3, C1, D2, E3, F2	No	-48.1	-9.1	-10.8	-4.0	14.1	11.0

CONFIRMATION

Following the experimentation and analysis of results, the confirmation experiments were run to demonstrate optimal performance at the chosen settings. In addition, since a main focus of the effort was to determine the setting for the LNA gain a series of five experiments were run for the gain values between the mid and high factors. It was determined that the optimal gain level for the system was at a value of mid+2 dB. This information will be used for the design specification of the LNA.

CONCLUSION

The basic conclusion from this effort is that the primary contributors to the effectiveness of the AIMS system were chosen properly. Blockage is the dominant detrimental factor. Cancellation is the dominant control for blockage. Engineering judgment and analysis suggest that an active signal canceler should be designed to provide the greatest amount of cancellation technologically possible. The receive filter affects system performance for both C/N and blockage. A narrower filter will result in improved performance. The receive filter, therefore, should be chosen to be as narrow as possible within the physical constraints levied on the system. The devices chosen both push the technological limits of canceler performance and the filter BW/size trade-off.

Results from the experiment are more interesting for the other parameters. First, an analysis of the effect of the receive filter did not match the expected result for C/N. Intuitively, a narrower filter should improve system performance against close-in nonblocking interferers. This effect should be seen in an improvement in C/N as the filter becomes narrower. A reverse effect is seen in the experimental data. Cancellation also seems to have little effect on C/N.

A closer look at the data and the construction of the noise matrix revealed the cause. The interferer frequencies chosen for the noise matrix were not close enough to the desired frequency to bring the broadband noise associated with the interferers into the radio channel and impact C/N. The effect seen was the increase in filter loss associated with narrower BW. Increased loss has an adverse effect on C/N. A more elaborate noise model would have given a clearer picture of the full effect of receive filter and canceler on overall system performance as measured by C/N.

Another unanticipated result was that the noise figure of the LNA has little effect on C/N and block count. Intuitively, a low noise figure would be chosen. However, this experiment shows that the LNA NF can be relaxed with little penalty on system performance. Detailed examination shows that it is the environment that drives the noise at the front end of the system not the LNA NF as in many other applications.

A more subtle effect on system performance is seen in the interaction between the system gain, set by the LNA gain and attenuator. System gain affects both the effectiveness against blocks and C/N. Therefore, a choice of LNA gain and attenuator to set this value must be made to yield a proper trade between C/N and blockage. Confirmation experiment data show the system gain optimum to be on the order of mid+2 dB. Most likely mid gain and high attention will be chosen. The rationale behind this will be to provide a safety margin of gain with no system performance degradation. System gain can then be adjusted to the desired value with the attenuator. A secondary advantage of higher values of attenuation will be a better VSWR match at the T/R filter input.

This case study is contributed by Glenn S. Davis and John Featherston, ITT Aerospace/Communications Division – USA. Thanks go to Dr. M. Phadke of Phadke Associates who assisted us throughout the conceptualization, development and analysis of this effort.

CHAPTER TWELVE 

Electronic Warfare Receiver System

ITT INDUSTRIES, AVIONICS DIVISION – USA

EXECUTIVE SUMMARY

BACKGROUND

To search for potential radar threats, a typical EW (electronic warfare) system employs one or more swept superheat (superheterodyne) receivers to perform signal detection and identification. Threats can be classified into two categories:

- Low probability of intercept radar (LPI) threats
- Pulse and pulse Doppler (PR) random threats

There are many different types of radar signals that are intended threats for a swept superheat receiver ranging from duty cycle pulsed to continuous wave (CW) waveform. The primary purpose of this project was to perform a set of experiments using Taguchi Methods™ to optimize the performance and robustness of a generic swept superheat receiver in the detection and identification of radar threats in a dense and time-varying EW environment.

A dramatic improvement of time to intercept (TTI) performance was seen by using optimum levels of control factors derived from a series of Taguchi experiments. The TTI criterion is based on the root-mean-squared (RMS) elapsed time of the swept superheat receiver before intercepting randomly selected radar threats. However, it does not provide clear indication of the receiver's dynamic performance in intercepting all radar threats as they appear randomly in the EW environment.

This project was planned using the probability of miss (Pm) to detect and identify radar threats versus elapsed time (T) as the performance characteristics, which is more effective in evaluating the receiver in a dynamic EW environment.

PROCESS

Instead of evaluating the system by a static approach, a dynamic approach was used to reduce the elapsed time to reach a desired level of detection of LPI threats. A high level simulator (Taguchi_New.For) written in FORTRAN was developed for performing matrix experiments on a generic swept superheat receiver. Six control factors were tested under seven noise factors which were generated by varying the number of LPI threats, starting time of each LPI threat for each set, signal amplitude of each LPI threat, starting time of each PR threat (first and second sets) and number of PR threats.

In this study, 18 experiments were conducted with a given set of control factors. The elapsed time (T) was chosen as the signal factor. The output response, probability of miss (Pm), was measured against the elapsed time (T) required by the swept superheat receiver to detect and identify all its intended LPI and PR threats.

BENEFITS

- Parameter design experiments using Taguchi Methods clearly resulted in a substantial improvement in performance and robustness of the threat detection and identification capability of a generic EW swept superheat receiver.
- Using the optimal set of control factors, the EW receiver demonstrated a remarkable 57% reduction of elapsed time required to reach a 95% probability of detection of all LPI threats in a dynamic EW environment while monitoring acceptable performance for detecting PR threats.
- Technical knowledge developed by this study will be applicable to future products.
- The new design is robust against many types of noise factors such as LPI threats and PR threats.
- ITT Avionics Division has developed a new, more effective and efficient way to evaluate and predict quality, reliability, and performance of the EW receiver system.
- Instead of spending mega dollars on upgrading hardware, it was possible to obtain a significant improvement by optimizing the software.

CASE STUDY

INTRODUCTION

ITT Avionics located in Clifton, New Jersey, is a leader in developing and manufacturing airborne electronic warfare (EW) systems for military applications. Major EW programs at ITT Avionics include AN/ALQ-172 (Pave Mint), AN/ALQ-165 (ASPJ), AN/ALQ-136 and Advanced Threat Radar Jammer (ATRJ).

In order to achieve high levels of ECM effectiveness, all EW systems developed by ITT Avionics perform threat signal detection identification functions. Typical receiver/processor architectures used for performing this function include time and frequency domain channelized receivers, compressive receivers, instantaneous frequency measurement (IFM) receivers, SAW and CCD based acoustic receivers, Bragg cell based optical receivers and swept superheterodyne (superheat) receivers.

Of special concern is the swept superheat, which was developed and used in many systems developed by ITT Avionics.

OBJECTIVES

To search for potential radar threats, a typical EW system employs one or more swept superheat receivers to perform signal detection and identification. The functional block diagram of a generic swept receiver is illustrated in Figure 1. By varying the frequency of the local oscillator (LO) to the mixer, the passband of the receiver is correspondingly stepped through the RF band of interest. Frequency isolation is achieved by feeding the IF signal through bandpass filters (BPFs). A narrowband (10 to 30 MHz) BPF is usually implemented for the signal detection and identification function. Parameters which affect the frequency scanning include the RF band limits, scanning rate, the BPF bandwidth, the scanning dwell time, which is the duration that the LO remains on the same frequency for signal detection, and the scanning strategy which predetermines the RF band priority and scanning sequences of the superheat swept receiver.

Figure 1: Functional Block Diagram of a Generic Swept Superheat Receiver

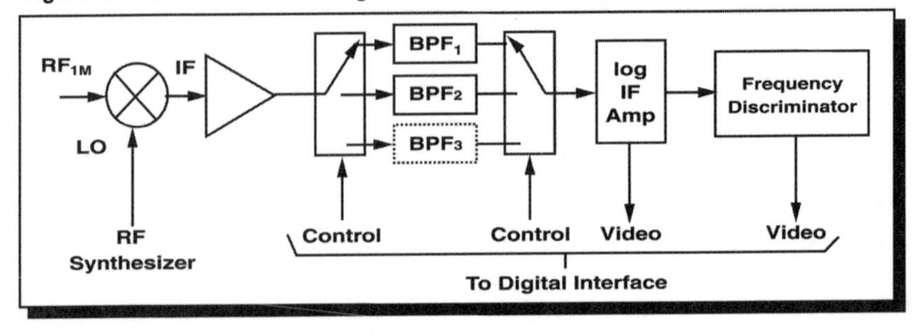

There are many different types of radar signals that are intended threats for a swept superheat receiver ranging from low duty cycle pulsed to continuous wave (CW) waveforms. To simplify this description, threats can be classified into two categories:

- Low probability of intercept radar (LPI) threats
- Pulse and pulse Doppler (PR) radar threats

The unique characteristics of an LPI threat are the relatively short pulse width (0.5 to 1.5 ms), short pulse repetition interval (PRI) (8 to 12 ms) and short time period available for interception by EW receivers. Because of this short period, the swept superheat receiver must intercept an LPI signal within several milliseconds of its first appearance in order to avoid potential disastrous consequences due to late detection.

With conventional PR radar threats, the radar signal pulse width is generally between 2 and 40 ms, the PRI ranges from 15 to 200 ms, and will usually remain in the EW environment for at least several seconds after its initial appearance.

The primary objective of this project was to perform a set of experiments using Taguchi Methods to optimize the performance and robustness of a generic swept superheat receiver in the detection and identification of radar threats in a dense and time-varying EW environment.

In a previous project in 1994, a series of (static) Taguchi experiments were conducted to optimize the threat detection and identification capability of a generic swept receiver. A computer simulator was developed to perform the Taguchi experiments. Using the optimum levels of control factors derived from these experiments, the receiver illustrated a dramatic improvement of time to intercept (TTI) performance when compared with the frequency scanning schemes currently implemented in ECM systems.

The TTI criterion is based on the root-mean-squared (RMS) elapsed time of the swept superheat receiver before intercepting a randomly selected radar threat. However, it does not provide clear indication of the receiver's dynamic performance in intercepting all radar threats as they appear randomly in the EW environment. Therefore a 1995 project was planned using the probability of miss (Pm) of detecting and identifying a radar threat versus the elapsed time (T) as the performance characteristic, which is more effective in evaluating the receiver in a dynamic EW environment.

Brainstorming

The brainstorming group included system, software and RF engineers who were thoroughly familiar with the design and programming techniques of a swept superheat receiver. The brainstorming effort focused on identifying the control and noise factors essential to the performance and robustness of a generic swept superheat receiver.

Ideal Function

One of the crucial processes in a robust design is to identify the ideal function for optimization. During the brainstorming and follow-up consultations, an ideal function was selected which measures the linear relationship between the elapsed time (T) of the threat when it appeared in the EW environment and the receiver's probability of miss (Pm) of detecting this threat. Derivation of this ideal function is described in the following paragraph.

For most EW receivers, the probability of detection (Pd) of a randomly selected radar threat increases exponentially as a function of T:

$$\mathbf{Pd = 1 - exp(-\beta T)}$$

where β is the time constant of the threat detection process and a function of the radar threat characteristics including the number and type of each radar threat and the frequency dwell requirements to support the post detection processing.

Let Pm = 1 – Pd be the probability of miss, then:

$$\mathbf{Pm = exp(-\beta T)}$$

and the ideal function is defined as:

$$\mathbf{y = ln\ (1/Pm) = ln\ [exp(\beta T)] = \beta T}$$

where a linear relationship exists between T and y.

P-Diagram

The factors studied in this project are listed below and are shown in Figure 2.

Control Factors:

1. Dwell time for LPI band scanning
2. Number of cycles for LPI band priority scanning
3. Dwell time for PR band scanning
4. Additional dwell time for LPI band after a threat detection
5. Additional dwell time for PR band after a threat detection
6. Receiver threshold level

Noise Factors:

1. Number of LPI threats
2. Starting time of each LPI threat (first set)
3. Starting time of each LPI threat (second set)
4. Signal amplitude of each LPI threat
5. Starting time of each PR threat (first set)
6. Starting time of each PR threat (second set)
7. Number of PR threats

Signal Factor:

Elapsed time (T)

The output response measured for optimization was the probability of miss (Pm) versus the elapsed time (T) required by the swept superheat receiver to detect and identify all its intended LPI and PR threats.

Figure 2: The P-Diagram of a Generic Swept Superheat Receiver/Controller

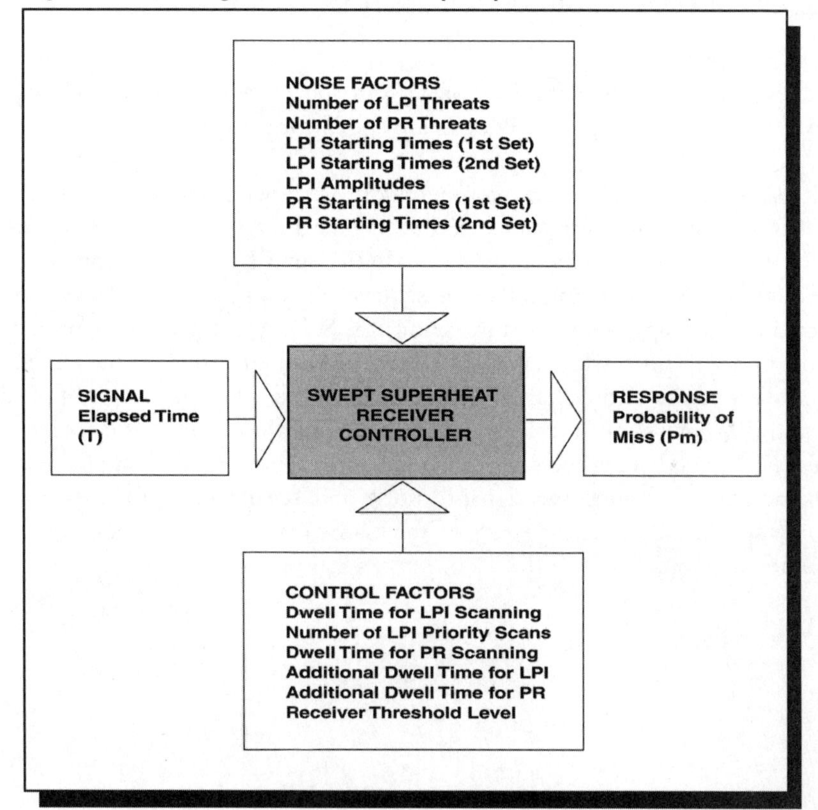

Simulator Development

As reported in 1994, a high level simulator (Taguchi_ New.For) written in FORTRAN was developed for performing matrix experiments on a generic swept superheat receiver. It simulates a typical ITT Avionics superheat receiver covering an RF range from 6 to 18 GHz and an IF frequency of 450 MHz. The receiver uses a single-frequency down conversion with its two IF passbands (upper and lower sidebands) located at ± 450 MHz from the LO. The bandwidth of the passband depends on the mode of operation of the swept receiver. For this project, only the narrowband mode (BPF = 10 to 30 MHz) scanning case was considered.

Features provided by the simulator and used in generating and collecting data in the Taguchi experiment included: (1) receiver frequency scanning sequence program, (2) frequency scanning strategy program, (3) radar signal generator for a dynamic EW environment (threat density varying from 250K to 1.5M pulses per second), (4) radar threat detection and identification, (5) receiver probability of miss (Pm) versus elapsed time (T) calculation and (6) data reduction processing as required by minute analysis.

DESCRIPTION OF TAGUCHI EXPERIMENTS

Test Array Setup

Tables 1 and 2 present the lists of factors and levels used in the $L_{18} \times L_{18}$ orthogonal array matrices. All levels are coded for proprietary reasons.

The $L_{18} \times L_{18}$ inner/outer array configuration, as shown in Figure 3, was used to conduct the matrix experiments. In the inner L_{18} array, six control factors, A to F, are housed from columns 2 to 7. In the outer L_{18} array, columns 1 and 2 are combined to accommodate the six levels of noise factor G data as collected via the upgraded software simulator. Subsequently, a series of data reduction and processing algorithms were performed including tracing and recording of the elapsed time (T) of individual radar threats as they applied to the EW environment, calculating and updating the receiver probability of miss (Pm), conducting the minute analysis of T versus Pm deriving the best-fit-fine and computing the signal-to-noise and sensitivity performance for each Taguchi experiment. Details of the matrix experiments are described in the next section.

Table 1: Control Factors and Levels

Symbol	Description	Factor Level
A	Dwell Time for LPI Scanning	Level 1 : a Level 2 : a + 9 Level 3 : a + 19
B	Number of LPI Priority Scans	Level 1 : b Level 2 : b + 4 Level 3 : b + 9
C	Dwell Time for PR Scanning	Level 1 : c Level 2 : c + 9 Level 3 : c + 39
D	Additional Dwell Time for LPI Threats	Level 1 : d Level 2 : d + 7 Level 3 : d + 15
E	Additional Dwell Time for PR Threats	Level 1 : e Level 2 : e + 7 Level 3 : e + 15
F	Receiver Threshold Level	Level 1 : f Level 2 : f + 0.1 Level 3 : f + 0.2

Table 2: Noise Factors and Levels

Symbol	Description	Factor Level
G	Number of LPI Threats	Level 1 : g Level 2 : g + 2 Level 3 : g + 4 Level 4 : g + 6 Level 5 : g + 8 Level 6 : g + 10
H	LPI Starting Times (First Set)	Level 1 : Time Set 1 Level 2 : Time Set 2 Level 3 : Time Set 3
I	LPI Starting Times (Second Set)	Level 1 : Time Set 4 Level 2 : Time Set 5 Level 3 : Time Set 6
J	LPI Amplitudes	Level 1 : Amp Set 1 Level 2 : Amp Set 2 Level 3 : Amp Set 3
K	PR Starting Times (First Set)	Level 1 : Time Set 7 Level 2 : Time Set 8 Level 3 : Time Set 9
L	PR Starting Times (Second Set)	Level 1 : Time Set 10 Level 2 : Time Set 11 Level 3 : Time Set 12
M	Number of PR Threats	Level 1 : m Level 2 : m + 5 Level 3 : m + 10

Figure 3: Test Configuration of Parameter Design Experiments

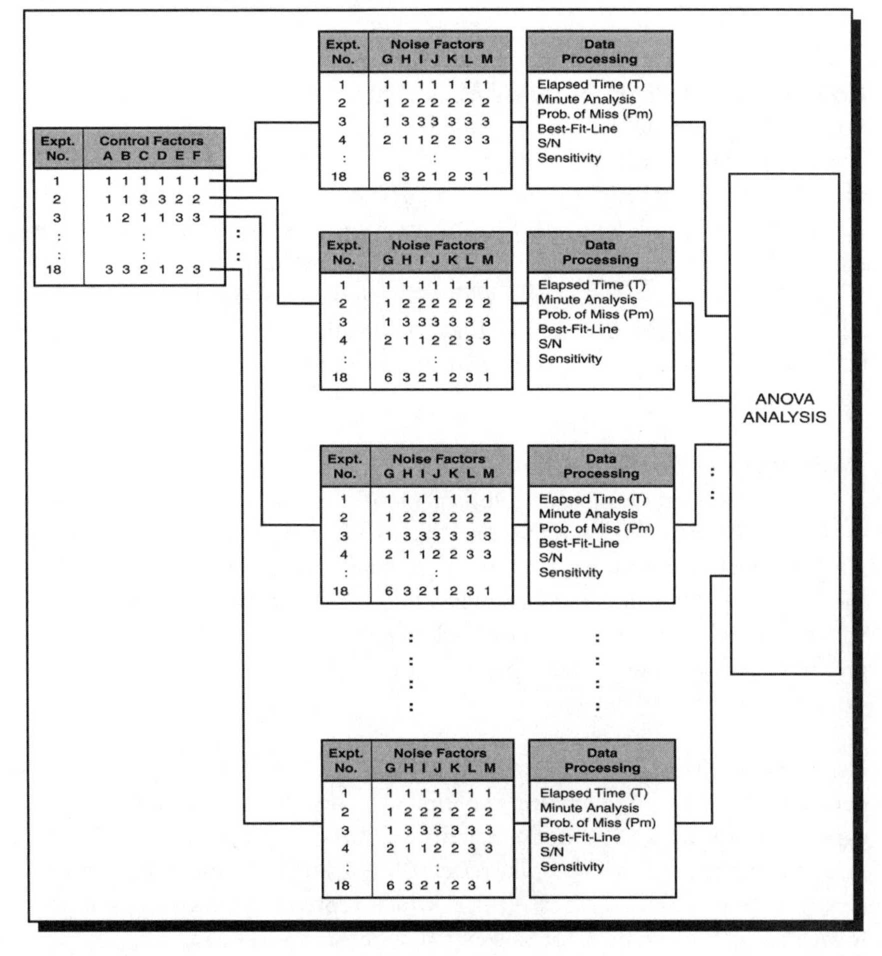

Parameter Design Experiments

A total of 18 Taguchi experiments were conducted with the different levels of control factors stated in Table 1. For each experiment, 18 outer array (matrix) experiments were performed representing the time-varying nature of the EW environment which was embedded in the noise factors in Table 2. The signal factor T was used to generate a dynamic spectrum to enable the measurement of the performance and robustness of the swept superheat receivers. Simulation data from each outer array experiment was collected and a series of data reduction and processing algorithms were then performed.

Calculation of Probability of Miss (Pm)

For each experiment, the elapsed time of each threat until its detection was continuously recorded. Elapsed times of radar threats (LPI or PR) were then placed in ascending order to calculate the receiver's probability of miss (Pm). As an example, if a total of 10 LPI threats were present and detected by the receiver with elapsed times of T_1 to T_{10}, the Pm would decrease with time as $1 - m/10$ at T_m, $m = 1, 2, \ldots, 10$ with m being the number of threats detected. To maintain finite values in the logarithmic calculation, Pm is set to $0.25/N$ when all of the expected threats (N) in the EW environment were detected.

Derivation of Best-Fit-Line

In order to obtain comprehensive insight of the relationship between the probability of miss (Pm) and the elapsed time (T) with a very limited data set, a minute analysis on the data set was performed. Elapsed times obtained in the computer simulation were quantized into 0.5-ms sectors. As a result, statistical dependence (log normal or Weibull distribution) of Pm versus T was removed, giving a much clearer presentation of the effects of control factors on the performance characteristic.

Using the results of minute analysis, individual curves of Pm versus T were generated for each of the matrix experiments. Simulation results of the first experiment of L_{18} array are illustrated in Figures 4 and 5 for LPI and PR threat detection. Each figure illustrates 18 individual curves of Pm (in dB) versus T. It also plots the best-fit-line, which represents the composite performance of Pm versus T for this experiment.

The slope of the best-fit-line is used to calculate the expected elapsed time for a specified Pm. Referring to Figure 4, since the slope (β) of the best-fit-line is 2.0809, therefore, in order for the receiver to detect at least 95% [or $-\log(\text{Pm}) = 13.01$ dB] of all the LPI threats, an expected elapsed time (T) is 6.252 ms.

Calculation of Signal-to-Noise Ratio and Sensitivity

After deriving the best-fit-line for each experiment, the signal-to-noise ratio (η) can be calculated. It is defined as the ratio of the slope of the best-fit-line to the average variance of the 18 response curves of the experiment. In addition the slope (β) of the best-fit-line was also computed, and is a strong indication of the sensitivity performance of the control factors. The values of η and β for the 18 experiments are shown in Table 3 for LPI and PR threat detection. Figures 4 and 5 show the result from No. 1 of inner array L_{18} for LPI and PR threats, respectively. An individual line represents one noise condition; therefore, there are 18 individual lines in each figure.

Figure 4: Probability of Miss vs. Elapsed Time for Experiment 1

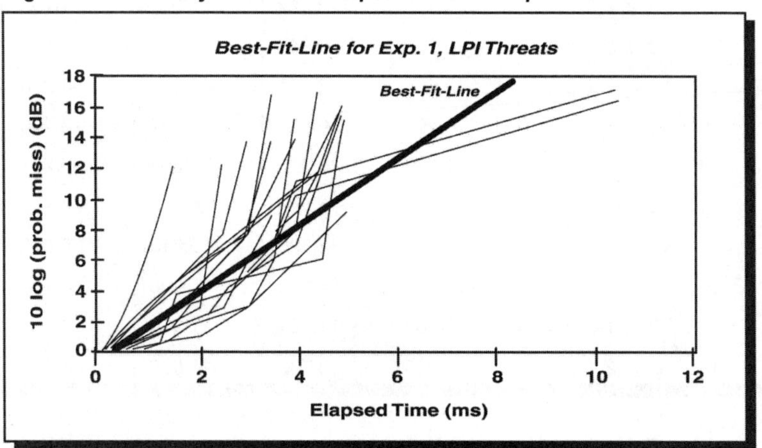

Figure 5: Probability of Miss vs. Elapsed Time for Experiment 1

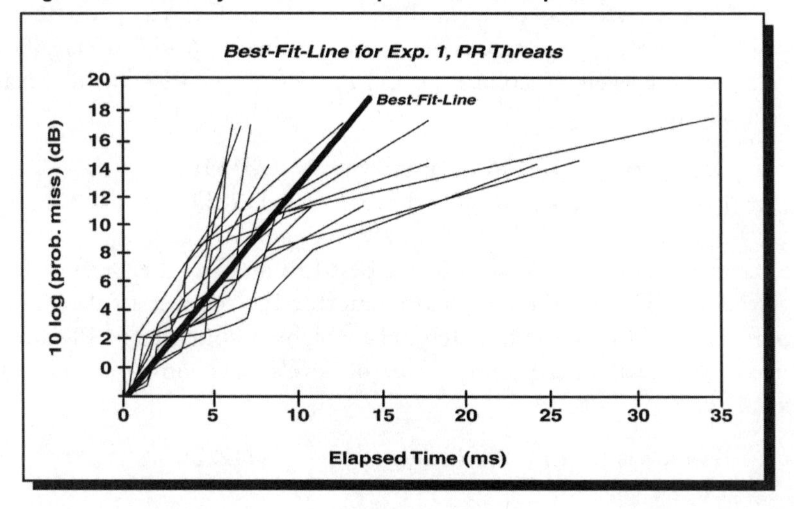

Table 3: Signal-to-Noise Ratios of Taguchi Experiments

Expt.	Control Factor	LPI Threat Detection		PR Threat Detection	
No.	A B C D E F	η (dB)	β (dB)	η (dB)	β (dB)
1	1 1 1 1 1 1	–2.128	6.365	–8.288	3.073
2	1 1 3 3 2 2	–7.019	1.607	–11.901	–4.261
3	1 2 1 1 3 3	–11.507	1.992	–11.373	2.062
4	1 2 2 2 2 2	–4.030	5.324	–5.710	4.356
5	1 3 2 2 1 1	–0.758	8.081	–13.224	–1.384
6	1 3 3 3 3 3	–11.298	0.358	–3.335	3.476
7	2 1 1 2 2 3	–12.159	4.303	–12.692	–1.883
8	2 1 2 3 1 3	–12.108	3.814	–10.165	–0.724
9	2 2 2 3 3 1	5.115	13.587	–10.011	–1.795
10	2 2 3 1 2 1	0.399	7.347	–6.570	–0.525
11	2 3 1 2 3 2	4.657	13.911	–15.019	–4.860
12	2 3 3 1 1 2	–0.451	6.952	–9.232	–2.466
13	3 1 2 1 3 2	1.276	8.905	–9.805	0.201
14	3 1 3 2 3 1	–1.774	6.449	–6.648	–0.940
15	3 2 1 2 1 2	2.882	12.235	–15.613	–5.026
16	3 2 3 2 1 3	–11.689	3.139	–10.408	–3.194
17	3 3 1 3 2 1	1.403	11.281	–17.998	–7.528
18	3 3 2 1 2 3	–11.916	3.722	-15.002	–4.353

Summary of Results

ANOVA Analysis

An ANOVA analysis was performed on the values of η and β for the 18 Taguchi experiments and results are shown in Figures 6 to 9. Using these results, the best levels of control factors for detecting LPI and PR threats respectively are:

1. Best detection for LPI radar threats: 231331
2. Best detection for PR radar threats: 112133

The control factor levels to achieve the best LPI and best PR performance are different. Due to the very short time for the receiver to detect and identify a LPI threat, it is considered to be a higher priority than a PR threat. In accordance with this priority, a set of levels of control factors were selected as:

3. Best levels for LPI/PR threats: 211222

Figure 6: η for LPI Threats

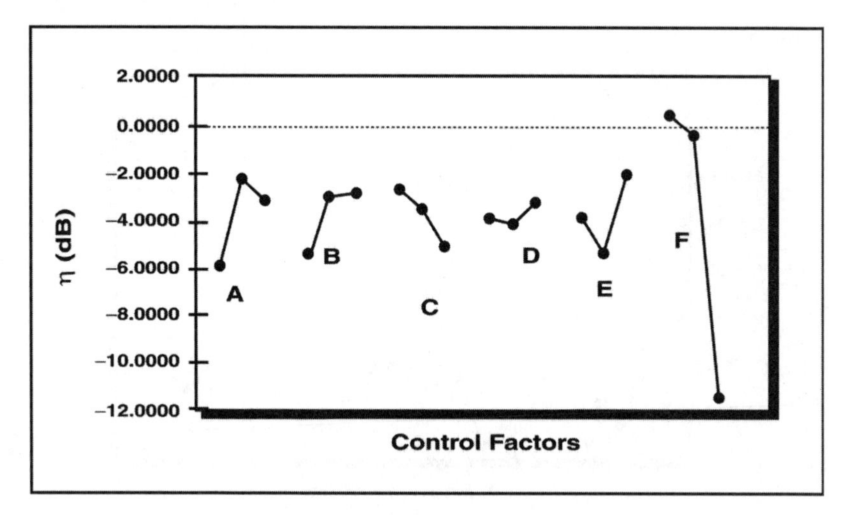

Figure 7: β for LPI Threats

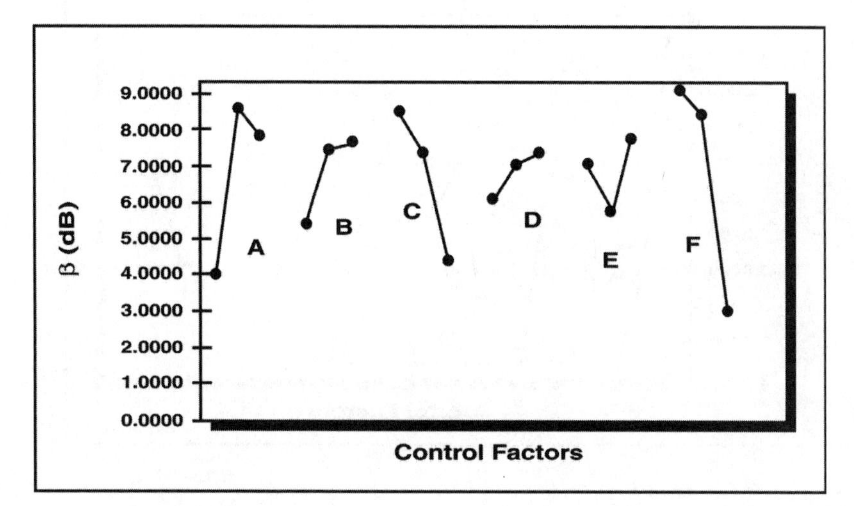

Figure 8: η for PR Threats

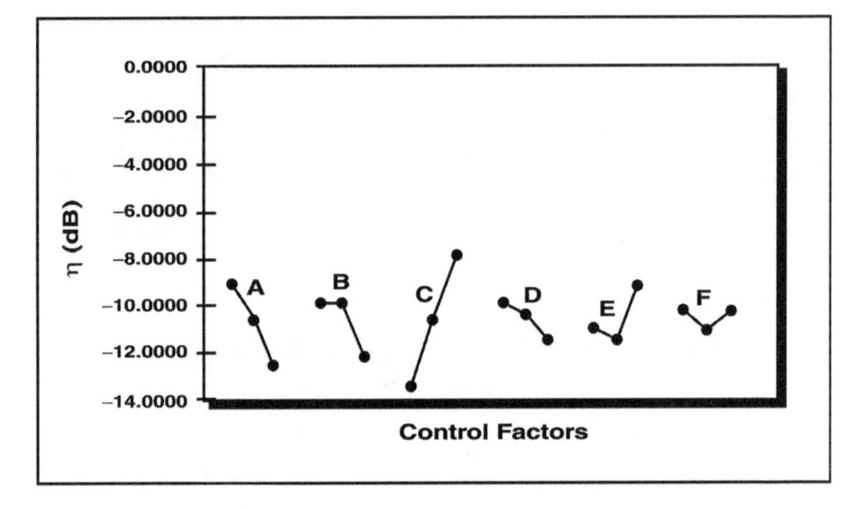

Figure 9: β for PR Threats

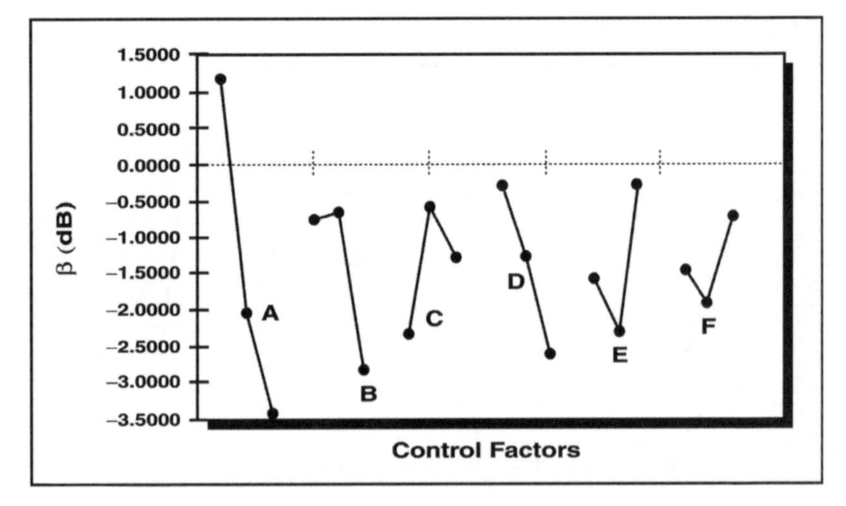

Confirmation Runs

Confirmation runs with control factor levels set at 231331 (best LPI), 112133 (best PR), 211222 (best LPI/PR), and 313112 (current baseline) were conducted. Results are summarized in Table 4.

Table 4: Results of Confirmation

Expt. No.	Control Factor ABCDEF	LPI Threat Detection		PR Threat Detection	
		η (dB)	β (dB)	η (dB)	β (dB)
Baseline	313112	–1.936	6.690	–7.380	–0.010
Best LPI	231331	4.869	15.344	–16.311	–6.538
Best PR	112133	–1.120	7.269	–4.692	4.890
Best LPI/PR	211222	6.980	13.911	–12.538	–1.412

Design improvements of the swept superheat receiver can be expressed in terms of Pm vs. T. The elapsed time (T) to reach a Pm of 5% (or a probability of detection of 95%) for different designs were calculated and listed in Table 5. The best LPI/PR design demonstrated a 57% reduction of elapsed time in LPI threat detection from 6.023 ms to 2.622 ms, while only increasing the PR threat detection from 13.026 ms to 15.306 ms of the current baseline design. Since PR threats have a long survival time (in tens of seconds) available for detection, this increment is practically insignificant in the performance of the swept superheat receiver.

Table 5: Time to Detect

Expt. No.	LPI Threat Detection Elapsed Time (ms)	PR Threat Detection Elapsed Time (ms)
Baseline	6.023	13.026
Best LPI	2.223	27.650
Best PR	5.635	7.409
Best LPI/PR	2.622	15.306

CONCLUSION

Parameter design experiments using Taguchi Methods have clearly resulted in a substantial improvement in performance and robustness of the threat detection and identification capability of a generic EW swept superheat receiver.

Since the timely detection of radar threats is a primary objective for an EW receiver, the criterion selected for optimization was the probability of miss (Pm) of detecting a radar threat versus the threat elapsed time (T).

Optimum control factor levels were chosen from the analysis of the results of the experiments. Using this set of control factors, the EW receiver demonstrated a *remarkable 57% reduction* of elapsed time required to reach a 95% probability of detection of all LPI threats in a dynamic EW environment while maintaining acceptable performance for detecting PR threats.

This case study is contributed by Paul Wang, ITT Industries, Avionics Division – USA. Special thanks to Dr. M. Phadke and Mel Kamenir for their guidance and valuable advice on this project.

CHAPTER THIRTEEN

Forward Error Correction Method for the MELP Vocoder

ITT INDUSTRIES – AEROSPACE/COMMUNICATIONS DIVISION
AND PHADKE ASSOCIATES, INC. – USA

EXECUTIVE SUMMARY

BACKGROUND

The mixed excitation linear predictive (MELP) algorithm was adapted in 1996 as the federal standard for the 2.4-kbps voice compression algorithm. This algorithm provides intelligible speech provided the channel noise is low. However, military communications systems often suffer from high channel noise. The resulting bit errors adversely affect the speech intelligibility. To overcome this limitation, forward error correction (FEC) techniques are selectively applied to the transmitted bits (speech parameters), in order to optimize robustness of speech intelligibility in military communication. The ITT Aerospace/Communications Division applied a parameter design to determine the sensitivity of the MELP speech parameters, and it used the results for the design of an FEC that best protects the speech parameters from channel errors.

PROCESS

The ideal function was defined by taking the magnitude of input speech as the input signal and the magnitude of synthesized speech (output) as the response. Noise factors were speaker sex, speech phrases, acoustic noise and bit error rate. Design parameters taken as control factors were jitter, gain indices, bandpass voicing, pitch index and others. Control factors were assigned to an L_{36} orthogonal array. The signal-to-noise ratio was used to evaluate the process, which was thus optimized for robustness of speech intelligibility.

BENEFITS

- The optimum design resulted in a 3-dB improvement in S/N ratio, which is equivalent to reducing the effective bit-error rate by a factor of two.
- The results were obtained in a time severalfold smaller than that needed for the traditional design approach (four to six days compared to four to six weeks).

- There was a substantial cost saving because money used to purchase poor measurement systems and time spent as a result of guesswork were avoided.
- Improvements were primarily possible because of the use of the dynamic S/N to objectively quantify speech intelligibility.
- The dynamic S/N developed in this research can be used not only for FEC optimization but also for developing alternative speech compression algorithms.

CASE STUDY

INTRODUCTION

The mixed excitation linear predictive (MELP) algorithm was adopted in 1996 as the federal standard for the 2.4-kbps voice compression algorithm. This algorithm provides intelligible speech provided the channel noise is low. However, military communications systems often suffer from high channel noise. The resulting bit errors adversely affect the speech intelligibility. To overcome this limitation, forward error correction (FEC) techniques are selectively applied to the transmitted bits (speech parameters).

This paper describes a parameter design experiment for determining the sensitivity of the MELP speech parameters, and the use of the results for designing a forward error correction (FEC) scheme which best protects the speech parameters from channel errors.

BACKGROUND

MELP Overview

MELP was developed by a team from Texas Instruments Corporate Research in Dallas and Atlanta Signal Processing, Inc. The vocoder is based on technology developed at the Center for Signal and Image Processing at the Georgia Institute of Technology in Atlanta. MELP was chosen by the U.S. Department of Defense (DoD) in 1996 as the new federal standard for the 2.4-kbps voice compression.

MELP is based on the linear predictive analysis of speech, similar to its predecessor, LPC-10e. The linear predictive model of speech formation has as its basis the simple fact that a speech sample can be approximated by a linear combination of its past samples. By minimizing the sum of the squared distance of the differences between the actual and predicted speech samples, a set of prediction coefficients for an autoregressive filter can be derived. This autoregressive filter essentially models the human vocal tract as a series of lossless tubes which are time varying in nature. Using this model, the speech is synthesized as the output of a linear, time-varying system, which is excited by a combination of quasi-periodic pulses and a random noise generator.

The functional block diagram of the MELP vocoder is shown in Figure 1. The vocoder consists of two independent sections, a speech analyzer and a speech synthesizer. Prior to analysis, the input speech is digitized by a 12-bit linear analog-to-digital (A/D) converter sampling at an 8 kHz rate resulting in an input data rate of 96,000 bits/s. The purpose of the speech analyzer is to perform bandpass voicing and pitch analysis, to calculate the coefficients for the autoregressive vocal tract filter, and to calculate the speech gain parameters. These parameters are bit packed into a 54-bit code word which is transmitted every 22.5 ms. The effective bit rate for the MELP bit stream is therefore 2400 bits/s.

Also shown in Figure 1 is the model for speech synthesis (known as speech production). The synthesizer receives the encoded speech parameters and attempts to reproduce the speech signal by using a mixed-excitation speech model. This model uses multiband mixing to simulate the frequency-dependent voicing strengths found in natural speech. The excitation generator mixes the periodic and random noise sources as an excitation source for the vocal tract filter. The primary effect of multiband mixed excitation is to reduce the buzz normally associated with LPC vocoders, thereby increasing speech intelligibility. The vocal tract effects in speech production are modeled after a time-varying linear autoregressive vocal tract filter. This filter improves the match between synthetic and natural speech waveforms, and results in a more natural quality of the speech output.

Figure 1: MELP System Block Diagram

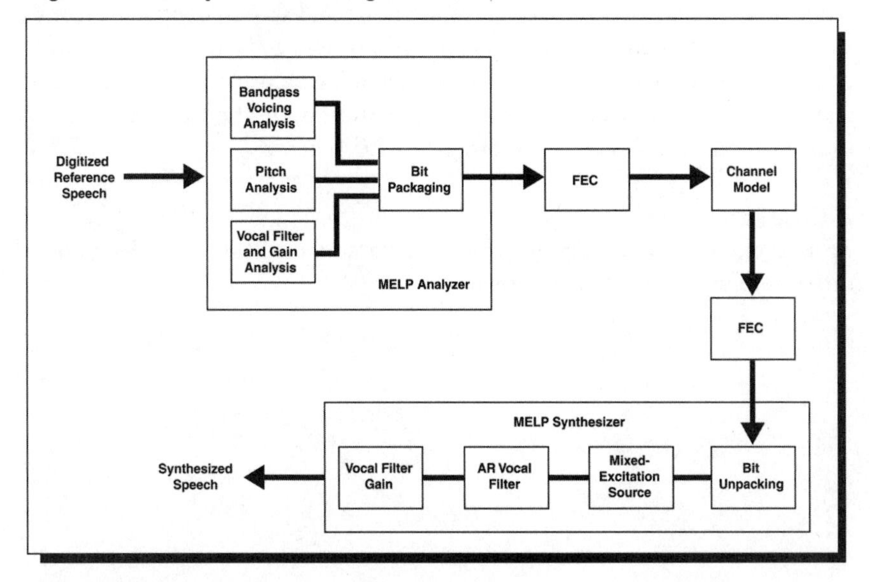

Forward Error Correction (FEC)

Figure 1 also shows the forward error correction processing and the channel model. The channel is modeled by a band-limited low pass filter with an additive white gaussian noise source. Forward error correction processing takes place before the MELP bit stream enters the channel and after it leaves the channel, prior to entering the MELP synthesizer. The purpose of the FEC is to add additional parity bits to the MELP parameters, thereby making them less sensitive to channel noise impairments. The MELP parameters are then corrected for channel bit errors prior to entering the MELP synthesizer.

Traditional Approach to FEC Design

The traditional approach to speech quality testing is based on a series of listener-scored tests known as the Diagnostic Rhyme Test (DRT) and the Mean Opinion Score (MOS). Both of these tests are performed by a jury system of listeners. These tests are expensive and time consuming, each costing thousands of dollars to run. A typical turnaround time for a DAM (diagnostic acceptance measure) or DRT score is in the order of four to six weeks. Finding an optimal FEC scheme that results in high speech quality with minimum bandwidth requires the speech quality tests to be repeated many times. This makes the traditional approach of designing the FEC scheme a highly subjective, time-consuming and expensive process.

Problem Formulation

The main function of the FEC is to use the available bandwidth to protect the MELP bits so as to render the best synthesized speech under a broad range of channel conditions, background noise and speech contents. Here we use the ideal function concept and the robustness S/N ratio to objectively quantify speech quality. The S/N ratio, being inexpensive to evaluate, facilitates rapid and reliable evaluation of the criticality of the different bits and thus promotes the design of an efficient FEC scheme.

P-Diagram

To facilitate the robust design process, a parameter diagram (P-diagram) of the system was generated as shown in Figure 2.

Figure 2: Parameter Diagram

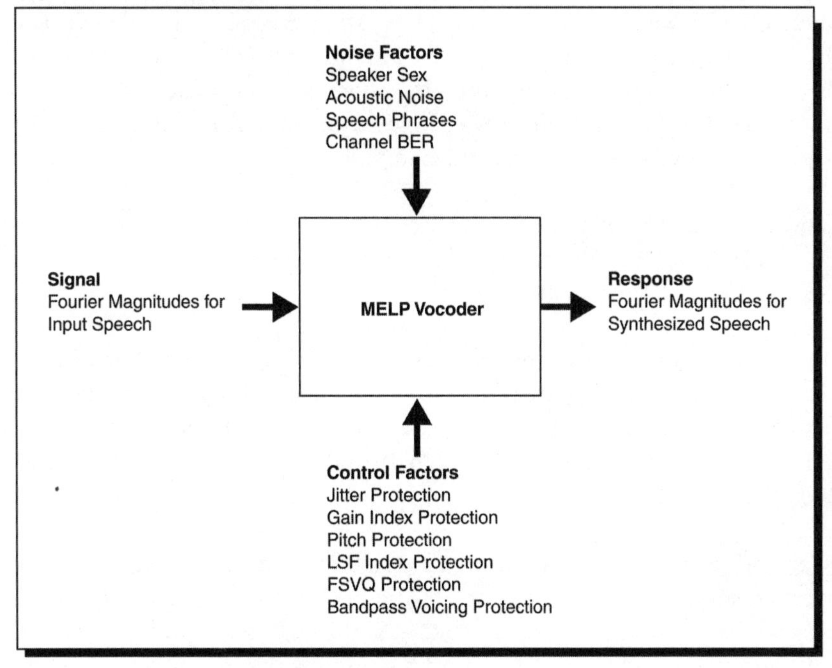

A description of the noise and control factors is given below.

Noise Factors

Speaker Sex: A variety of speakers, both male and female, were used in the experiments.

Speech Phrase(s): A variety of phrases and words, derived from standardized speech DAM and DRT testing, were used in the experiments.

Acoustic Noise: A variety of acoustic background noise sources, including a quiet background, a vehicular background and a helicopter background, were used in the experiments

Bit Error Rate: To represent the channel impairments, the experiments were conducted with a 1%, 5%, 10% and 20% channel bit error rate (BER).

Control Factors

Jitter: The jitter flag is set to 1 to indicate that the pulse component of the mixed excitation should be aperiodic rather than purely periodic. Jitter is set to 25% if this bit is set to 1 and is 0% otherwise.

Gain 1 Index: The input speech gain is measured twice per frame using a pitch-adaptive windowing procedure. The estimated gain is the RMS value of the input signal over the window. The first window is centered at the mid-point of the current frame and the second window is centered at the end of the frame. The first gain index is encoded using an adaptive 3-bit algorithm whose level is relative to the second gain index.

Gain 2 Index: The window for the second gain index is centered at the end of the current frame. The second gain index is quantized with a 5-bit uniform quantizer ranging from 10 to 77 dB.

Bandpass Voicing: The bandpass voicing index is quantized to 0 or 1 in each band to indicate the presence of a voice or an unvoiced signal in the passband frequency range. The analysis filters are sixth order Butterworth filters with the following passbands: 0–500 Hz, 500–1000 Hz, 1000–2000 Hz, 2000–3000 Hz and 3000–4000 Hz.

Pitch Index: Pitch is quantized on a logarithmic scale using a 99-level quantizer ranging from 20 to 160 samples. The pitch values are mapped into a 7-bit code word using a lookup table. An all-zero code word indicates an unvoiced frame and all 28 codes with a hamming distance weight of 1 or 2 are reserved to indicate unvoiced frames and frame erasures. It represents the fundamental frequency for the excitation signal.

LSFI – Stages 1 to 4: The line spectral frequency index for stage 1 through stage 4. The LPC coefficients are converted into LSFs and quantized using a multistage vector quantization of the LSFs. The MSVQ code book has four stages with indices of 7, 6, 6 and 6 bits each. The quantized LSF vector is the sum of these four vectors, one vector from each stage.

FSVQ Index: The 10 Fourier magnitudes are encoded using an 8-bit vector quantizer. The index of the transmitted code vector is chosen to minimize the Euclidean distance between the input and the code vector. The FSVQ index, in conjunction with the LSF indices, form the basis for the vocal tract autoregressive filter.

Response Variable

Speech Magnitude: The speech Fourier coefficient magnitudes are calculated by performing a short-term, discrete Fourier transform of the speech. For the robust design experiments, the Fourier magnitude was calculated only during a speech phrase and typically lasted on the order of 0.2 s to 0.5 s. The speech Fourier coefficient magnitude was calculated for both the reference speech and the synthesized speech waveforms.

Ideal Function

Two approaches could be used to define the ideal function of the total MELP system (which includes the FEC): time domain and frequency domain. First we started with the time domain approach to quickly discover that the resulting S/N ratio had little correlation with the speech quality perceived by a listener. So, we discarded the approach.

In frequency domain, the ideal function can be defined as follows: the Fourier coefficient magnitudes of the synthesized MELP speech should be proportional to the Fourier coefficient magnitudes of the input reference speech. There was some concern about how to handle the effect of the phases of the different Fourier coefficients. A literature search indicated that the human ear is insensitive to phase variations of sounds. Over a sufficient number of cycles, a pure sine wave tone sounds the same whether it starts at 0 degrees, 45 degrees, 90 degrees or any other arbitrary starting phase. Thus, the reference speech and the synthesized speech may be out of phase and yet still sound the same. Further, MELP does not transmit the pitch phase as part of its parameters, therefore, the synthesizer always begins the periodic excitation source at the same phase regardless of the phase of the reference speech waveform.

Figure 3 shows the plot of Fourier coefficient magnitudes of the MELP synthesized speech vs. those for the reference input speech. A perfect speech reproduction would give us a straight line through the origin with no dispersion. However, since MELP is a lossy compression scheme, the synthesized

Figure 3: Ideal Function

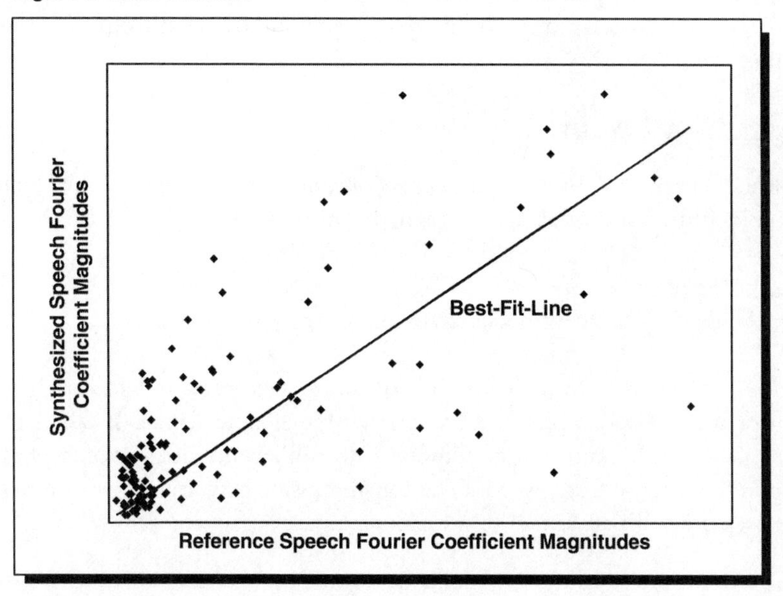

speech will deviate somewhat from the reference speech. This deviation becomes greater as the MELP parameters are corrupted by channel errors, which results in a larger variance from the ideal straight line. The overall effect is a decreasing S/N ratio as the speech quality is degraded by channel errors.

We validated the ability of the frequency domain–based S/N ratio to quantify the speech quality through informal speech tests in the ITT speech lab. These results showed a very good correlation between the S/N ratio and the DAM/DRT scores. Other researchers have also reported very good results using frequency domain-based objective speech quality testing, including K. H. Lam et al. [2]. However, their metric is very different from the S/N ratio developed here.

Signal-to-Noise Ratio

The appropriate signal-to-noise ratio for this problem is the continuous-to-continuous type S/N ratio that is also known as the zero-point proportional type S/N ratio. Let β be the slope of the best-fit-line through the origin to the Fourier coefficient magnitude data, as shown in Figure 3. Also, let σ^2 be the variance of the data around the best-fit-line. Then, the signal-to-noise ratio, denoted by η, is defined by the following equation:

$$\eta = 10 \log (\beta^2/\sigma^2)$$

Simulator Development

The simulation experiments were performed by modifying the DoD standard MELP algorithm developed jointly by Texas Instruments and Atlanta Signal Processing Incorporated to include the effects of noisy channels and FEC. The speech analysis and synthesis routines were unchanged from the original source code.

Experiment Layout

Through a series of brainstorming sessions, noise factors and control factors were identified and levels were assigned to these factors.

Noise Factor Levels

Levels for the first three noise factors are shown in Table 1.

Since there were a limited number of noise factors, a full factorial noise source was created, which used a variety of male and female speakers in all three acoustic background conditions with both continuous speech phrases and individual (discrete) words. The fourth noise factor was bit error rate and we used a 5% bit error rate in our simulation.

Table 1: Noise Factor Levels

Noise Factors	Level 1	Level 2	Level 3
Speaker Sex	Male	Female	----
Speech Phrase	Continuous	Discrete	----
Acoustic Noise Background	Quiet	Vehicular	Helicopter

Control Factor Levels

Table 2 indicates the control factors and the levels used in this experiment. Each of the levels indicates the amount of FEC which was applied to a particular parameter. For instance, light indicates that none of the parameter bits were protected, medium indicates that the two MSBs (most significant bits) were protected, while heavy indicates that all the bits were protected by FEC.

Table 2: Control Factor Levels

Control Factors	Level 1	Level 2	Level 3
Jitter	Light	Heavy	---
Gain 1 Index	Light	Medium	Heavy
Gain 2 Index	Light	Medium	Heavy
Bandpass Voicing	Light	Medium	Heavy
LSFI – Stage 1	Light	Medium	Heavy
LSFI – Stage 2	Light	Medium	Heavy
LSFI – Stage 3	Light	Medium	Heavy
LSFI – Stage 4	Light	Medium	Heavy
Pitch	Light	Medium	Heavy
FSVQ Index	Light	Medium	Heavy

Experiment Design

Based on the identified control factors, an L_{36} orthogonal array was chosen for the test configuration for the robust design experiments. Since there is one two-level factor and nine three-level factors, a $2^{11} \times 3^{12}$ L_{36} array was chosen to hold the MELP parameters and was modified as follows. The first column in the standard L_{36} array was used to house the two-level control factor and columns 12 through 20 were used to house the three-level control factors. Columns 2 through 11 and 21 through 23 were not used in the experiment design. All subsequent calculations were modified as appropriate to take into account the missing columns in the experiment design.

The control orthogonal array for the experiments is shown in Table 3. Due to the competition-sensitive nature of this design, the 10 control parameters have been coded and are not indicated by name in the table; rather they are listed by alphabetic letter (e.g. A, B, C, D, E, F, G, H, I and J).

EXPERIMENTAL RESULTS AND ANALYSIS

Experimental Results

The results of the L_{36} experiments, namely the S/N ratios and sensitivity, are shown in Table 4. The experiments were conducted with 5% BER.

Table 3: Control Factors Orthogonal Array

Expt. No.	A: Col. 1	B: Col. 12	C: Col. 13	D: Col. 14	E: Col. 15	F: Col. 16	G: Col. 17	H: Col. 18	I: Col. 19	J: Col. 20
1	1: Light	1: Light	1: Light	1: Light	1: Light	1: Light	1: Light	1: Light	1: Light	1: Light
2	1: Light	2: Med.	2: Med.	2: Med.	2: Med.	2: Med.	2: Med.	2: Med.	2: Med.	2: Med.
3	1: Light	3: Heavy	3: Heavy	3: Heavy	3: Heavy	3: Heavy	3: Heavy	3: Heavy	3: Heavy	3: Heavy
4	1: Light	1: Light	1: Light	1: Light	1: Light	2: Med.	2: Med.	2: Med.	2: Med.	3: Heavy
5	1: Light	2: Med.	2: Med.	2: Med.	2: Med.	3: Heavy	3: Heavy	3: Heavy	3: Heavy	1: Light
6	1: Light	3: Heavy	3: Heavy	3: Heavy	1: Light	1: Light	1: Light	1: Light	1: Light	2: Med.
7	1: Light	1: Light	1: Light	2: Med.	3: Heavy	1: Light	2: Med.	3: Heavy	3: Heavy	1: Light
8	1: Light	2: Med.	2: Med.	3: Heavy	1: Light	2: Med.	3: Heavy	1: Light	1: Light	2: Med.
9	1: Light	3: Heavy	3: Heavy	1: Light	2: Med.	3: Heavy	1: Light	2: Med.	2: Med.	3: Heavy
10	1: Light	1: Light	1: Light	3: Heavy	2: Med.	1: Light	3: Heavy	2: Med.	3: Heavy	2: Med.
11	1: Light	2: Med.	2: Med.	1: Light	3: Heavy	2: Med.	1: Light	3: Heavy	1: Light	3: Heavy
12	1: Light	3: Heavy	3: Heavy	2: Med.	1: Light	3: Heavy	2: Med.	1: Light	2: Med.	1: Light
13	1: Light	1: Light	2: Med.	3: Heavy	1: Light	3: Heavy	2: Med.	1: Light	3: Heavy	3: Heavy
14	1: Light	2: Med.	3: Heavy	1: Light	2: Med.	1: Light	3: Heavy	2: Med.	1: Light	1: Light
15	1: Light	3: Heavy	1: Light	2: Med.	3: Heavy	2: Med.	1: Light	3: Heavy	2: Med.	2: Med.
16	1: Light	1: Light	2: Med.	3: Heavy	2: Med.	1: Light	1: Light	3: Heavy	2: Med.	3: Heavy
17	1: Light	2: Med.	3: Heavy	1: Light	3: Heavy	2: Med.	2: Med.	1: Light	3: Heavy	1: Light
18	1: Light	3: Heavy	1: Light	2: Med.	1: Light	3: Heavy	3: Heavy	2: Med.	1: Light	2: Med.
19	2: Med.	1: Light	2: Med.	1: Light	3: Heavy	3: Heavy	3: Heavy	1: Light	2: Med.	2: Med.
20	2: Med.	2: Med.	3: Heavy	2: Med.	1: Light	1: Light	1: Light	2: Med.	3: Heavy	3: Heavy
21	2: Med.	3: Heavy	1: Light	3: Heavy	2: Med.	2: Med.	2: Med.	3: Heavy	1: Light	1: Light
22	2: Med.	1: Light	2: Med.	2: Med.	3: Heavy	3: Heavy	1: Light	2: Med.	1: Light	1: Light
23	2: Med.	2: Med.	3: Heavy	3: Heavy	1: Light	1: Light	2: Med.	3: Heavy	2: Med.	2: Med.
24	2: Med.	3: Heavy	1: Light	1: Light	2: Med.	2: Med.	3: Heavy	1: Light	3: Heavy	3: Heavy
25	2: Med.	1: Light	3: Heavy	2: Med.	1: Light	2: Med.	3: Heavy	3: Heavy	1: Light	3: Heavy
26	2: Med.	2: Med.	1: Light	3: Heavy	2: Med.	3: Heavy	1: Light	1: Light	2: Med.	1: Light
27	2: Med.	3: Heavy	2: Med.	1: Light	3: Heavy	1: Light	2: Med.	2: Med.	3: Heavy	2: Med.
28	2: Med.	1: Light	3: Heavy	2: Med.	2: Med.	2: Med.	1: Light	1: Light	3: Heavy	2: Med.
29	2: Med.	2: Med.	1: Light	3: Heavy	3: Heavy	3: Heavy	2: Med.	2: Med.	1: Light	3: Heavy
30	2: Med.	3: Heavy	2: Med.	1: Light	1: Light	1: Light	3: Heavy	3: Heavy	2: Med.	1: Light
31	2: Med.	1: Light	3: Heavy	3: Heavy	3: Heavy	2: Med.	3: Heavy	2: Med.	2: Med.	1: Light
32	2: Med.	2: Med.	1: Light	1: Light	1: Light	3: Heavy	1: Light	3: Heavy	3: Heavy	2: Med.
33	2: Med.	3: Heavy	2: Med.	2: Med.	2: Med.	1: Light	2: Med.	1: Light	1: Light	3: Heavy
34	2: Med.	1: Light	3: Heavy	1: Light	2: Med.	3: Heavy	2: Med.	3: Heavy	1: Light	2: Med.
35	2: Med.	2: Med.	1: Light	2: Med.	3: Heavy	1: Light	3: Heavy	1: Light	2: Med.	3: Heavy
36	2: Med.	3: Heavy	2: Med.	3: Heavy	1: Light	2: Med.	1: Light	2: Med.	3: Heavy	1: Light

Factor Effects

Factor effect plots were generated for the sensitivity and S/N ratio data given in Table 4. Figure 4 shows the factor effect plots for the S/N ratio. Note that for optimization of speech quality, only the analysis of the S/N ratio is relevant. Sensitivity does not play a role.

Table 4: S/N Ratio and Sensitivity for 5% BER

Expt. No.	S/N Ratio (dB) @ 5% BER	Sensitivity @ 5% BER
1	–96.5	0.75
2	–95.7	0.74
3	–92.6	0.85
4	–95.8	0.74
5	–94.3	0.85
6	–94.2	0.81
7	–96.2	0.74
8	–94.0	0.87
9	–94.6	0.80
10	–94.7	0.81
11	–95.7	0.75
12	–94.2	0.81
13	–94.0	0.80
14	–95.3	0.79
15	–95.3	0.77
16	–94.1	0.82
17	–95.3	0.81
18	–93.9	0.80
19	–94.8	0.79
20	–95.2	0.86
21	–93.1	0.83
22	–95.1	0.75
23	–94.2	0.85
24	–95.0	0.77
25	–94.8	0.77
26	–93.2	0.83
27	–95.0	0.77
28	–94.8	0.82
29	–92.7	0.88
30	–95.0	0.77
31	–93.6	0.84
32	–94.9	0.82
33	–94.6	0.78
34	–95.8	0.75
35	–94.4	0.83
36	–93.5	0.82

It is clear from Figure 4 that factor D has the greatest effect on the S/N ratio and hence will have the greatest effect on speech quality. Factors B and F have the next largest effect, whereas factors A, G and J have a smaller effect. The remaining factors have only minor contributions to the S/N ratio. In the next section we use this rank ordering of the factor effects in designing an optimal FEC scheme.

Figure 4: Factor Effects for the S/N Ratio

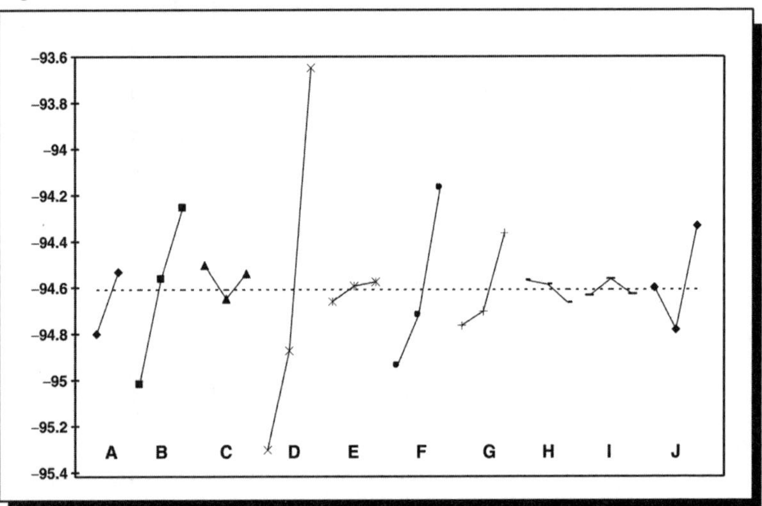

FEC Design and Confirmation

FEC Selection Constraints

Based on the results obtained from the Taguchi experiments and the factor effect plots, a final FEC configuration can be chosen and its performance verified. Final selection of an optimal FEC scheme is an exercise in linear integer programming techniques and must balance the trade-offs between speech quality versus available channel bandwidth. In a typical system, the number of parameters that may be protected by an FEC scheme is constrained by the available channel bandwidth. For a given BER, the FEC schemes are constrained to a maximum number of bits that may be transmitted without exceeding the available channel bandwidth.

FEC Design

For the system under consideration, the FEC design is constrained by the available channel bandwidth to a maximum number of 14 protected bits. In addition, the channel BER was specified to be 5%. Given these two constraints, all seven possible FEC schemes were investigated, ranging from

7 protected bits up to 14 protected bits. The objective of this design process is to produce the FEC scheme which will produce the best speech quality, and hence the highest S/N ratio.

The factors effect plots for 5% BER were used to select the best combination of settings for these seven FEC schemes. Table 5 summarizes the factor settings for each of the candidate approaches. This table also shows the predicted S/N ratio and observed S/N ratio for each of the candidate FEC schemes.

Table 5: FEC Protection Scheme Prediction Summary

FEC Scheme	Factor Settings	No. of Protected Bits	Predicted S/N Ratio (dB)	Observed S/N Ratio (dB)
Baseline	A1, B1, C1, D1, E1, F1, G1, H1, I1, J1	0	---	–96.5
1	A1, B1, C1, D3, E1, F1, G1, H1, I1, J1	7	–94.95	–94.84
2	A1, B2, C1, D2, E1, F2, G1, H1, I1, J1	7	–94.38	–94.94
3	A1, B3, C1, D3, E1, F1, G1, H1, I1, J1	12	–93.66	–94.14
4	A2, B2, C1, D3, E1, F2, G1, H1, I1, J1	12	–93.91	–93.74
5	A2, B3, C1, D3, E1, F1, G1, H1, I1, J1	13	–94.13	–93.74
6	A1, B1, C1, D3, E1, F3, G1, H1, I1, J1	14	–93.72	–94.04
7	A1, B3, C1, D3, E1, F2, G1, H1, I1, J1	14	–93.68	–93.84

Discussion of Results

The data in Table 5 reveal some interesting insights into FEC design. An examination of the FEC schemes shows that the two 12-bit schemes are predicted to perform well against channel errors. The control factors for these two schemes indicate two differing approaches to FEC protection.

Scheme 3 protects parameters B and D heavily, while scheme 4 protects parameter D heavily and parameters A, B and F moderately. The difference in these two approaches illustrates an interesting problem in FEC design. The paramount question facing the designer is, Given a limited number of bits that may be protected by FEC, what is the best way to allocate the FEC function to provide the best overall speech quality? To solve this problem, two differing schools of thought have prevailed.

One approach to designing an FEC scheme is to provide heavy protection on only a few parameters. The second approach is to provide moderate protection on as many parameters as possible. The total number of bits that may be protected limits both of these techniques. The results from the robust design experiments indicate that scheme 3 has the lowest predicted S/N ratio with a value of −93.66 dB while scheme 4 has a predicted S/N ratio of −93.91 dB. These values compare favorably with the 13- and 14-bit schemes, whose S/N ratios range from −93.68 dB to −94.13 dB.

The observed S/N ratio shows that schemes 4 and 5 perform the best, while scheme 3 does not perform as well as predicted. An examination of the configurations reveals that FEC schemes 4 and 5 are very similar and differ in only one of the FEC protected parameters. Scheme 3, however, differs in two protected parameters (e.g., A and F), and in the levels of protection for these parameters. One conclusion from the confirmation results is that for the best performance, control factor D should be covered heavily, control factors A, B and F should be covered nominally and the remaining factors should be covered lightly.

The resulting FEC design is an interesting combination of the two differing FEC ideologies previously discussed. It is actually a hybrid of heavily, moderately and lightly protected parameters. This is a result that would not be expected using conventional design techniques. It is only through robust design methods that this FEC configuration was investigated and chosen. Using this FEC scheme, the designer can expect a 2.76-dB improvement in the S/N ratio over the baseline FEC approach. These results were confirmed through informal listening tests at ITT's speech lab. As a final recommendation, schemes 4 and 5 should undergo formal DAM and DRT scoring for a final confirmation of these results.

Conclusion

In conclusion, the objective voice scoring methods developed under this project, in conjunction with robust design techniques and the Taguchi Methods, have proved to be a cost-effective approach to designing an FEC scheme for the MELP vocoder. Using these techniques, an improvement of 2.76 dB was observed over the baseline FEC approach. The analysis in this paper demonstrates the importance of applying robust design techniques early in the design process. It also shows how a small number of Taguchi experiments can be used to evaluate competing FEC schemes in a cost-effective and timely manner.

This case study is contributed by Edward Wojciechowski of ITT Industries – Aerospace/Communications Division and Madhav S. Phadke of Phadke Associates, Inc. The authors wish to thank Matthew Petrulli for his assistance in modifying the MELP simulator for this experiment.

CHAPTER FOURTEEN 

Robust Design of an Electrical Insulator

ITT NIGHT VISION AND ITT DEFENSE AND ELECTRONICS – USA

EXECUTIVE SUMMARY

BACKGROUND

This study addresses the improvement of an encapsulation process used in electro-optical image intensifiers to isolate the electrically active surfaces from ambient influences. The need for improvement came from customer-imposed environmental test requirements which were not being consistently met in the production environment, resulting in customer dissatisfaction and increased costs due to rework and scrap.

The customer-imposed environmental testing simulated the system's end user's conditions. The system consists of this electro-optical image intensifier, housing and associated power supply subsystems.

PROCESS

The approach selected for this improvement effort was product and process parameter design optimization using Dr. Taguchi's methods. The figure of merit used to select the optimum parameter set was a double-signal dynamic signal-to-noise ratio. This engineering metric replaced the customer's quality level metric called "luminance gain" typically measured using the image intensifier module. At the customer's quality level, constant "luminance gain" was ideal, but this characteristic was impacted by many other parameters at the product level, so it was necessary to isolate the electrical encapsulation process and measure "coupons" designed to simulate the function of the encapsulant, electrical insulation.

An orthogonal array was used to efficiently study eight process and product control parameters against two noise parameters, one environmental and the other a manufacturing condition. The orthogonal array allowed evaluation of 4,374 possible control parameter combinations using only 18.

Benefits

As a result of the improvement effort, the following benefits were realized:

- Scrap and rework were eliminated.
- The "yield" associated with the customer's environmental testing went to 99.99%.
- Customer satisfaction improved as a result, reducing the number and frequency of sampling and screening test requirements (the test duration was reduced from ten days to five days and the number of samples was reduced from 13 to 3).
- The optimum process and product design was applied to other product types with complete success, requiring only minor tuning.
- Management and engineering resources could be shifted from "fire-fighting" to improving other product and process designs.
- The production problem had been addressed for 18 months using traditional methods without success. This effort required less than one month for experimental execution to production line implementation.
- Although an exact value for monetary savings was not calculated, it is reasonable to estimate that waste reduction over the years since implementation exceeds $1M.

CASE STUDY

Introduction

Designing highly reliable products and the capable processes to produce them using the simplest mechanisms at the lowest cost is the primary function of the engineer. Reliability can be defined as the ability of the product (the engineered system) to maintain its function over its intended lifetime while exposed to undesirable external factors during usage, as shown in Figure 1. Dr. Genichi Taguchi developed a method of design engineering that focuses on developing this economical robust technology from which many families of products may evolve, avoiding the need to develop each product individually. This approach reduces product development cycle time and decreases downstream manufacturing costs by reducing product design and processing sensitivity to undesirable external factors. This method, called "robust design," can be used for the development of technology, products, processes and measurement systems.

The method offers an alternative approach to extensive product life testing typically used for developing highly reliable products. Rather than using the product in accelerated testing, exposing it to the repeated impact of external factors responsible for performance degradation, an alternative approach is used to seek selection of the levels of controllable factors which bring the

Figure 1: *Reliability as Endurance of Function*

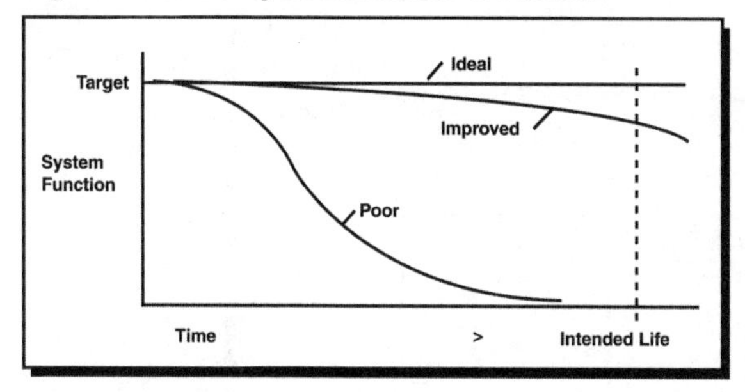

engineered system closer to its ideal function while it is exposed to usage and unwanted external factors.

This study used Dr. Taguchi's dynamic parameter design approach to improve the reliability of a night vision viewing product line with regard to environmentally induced performance degradation. The following sections present a parameter design effort focused on improving the reliability of the image sensor module used in that night viewing product line. The module is used in many products produced at ITT Night Vision.

Figure 2

Figure 3

ITT Electro-Optical Products Division, located in Roanoke, Virginia, maintains the world's largest Generation III night vision manufacturing facility, employing approximately 700 people. Primarily a Department of Defense contractor, EOPD fabricates night vision intensification systems, including "goggles," which enable pilots and ground troops to navigate during dark hours for tactical advantage. Figure 2 shows a PVS (personal viewing system) which has one sensor module located behind the objective lens opposite the eyepieces.

At the heart of the module, shown in Figure 3, is an image intensifier tube, shown in Figure 4, whose outer surface has exposed electrical elements which must be covered to provide environmental isolation. In addition, the material covering the tube's surface must electrically isolate the exposed metallic elements. The space allowed for the encapsulant material is very small in order to reduce weight and size, adding to the challenge for the process in introducing the material. A paramount constraint to the engineered system is the requirement that the image tube not be exposed to greater than 95^{0}C in order to maintain the tube's ability to amplify low levels of input light by a constant value which is stable over time (stable luminous gain).

Figure 4: Cross Section of the Image Sensor Module

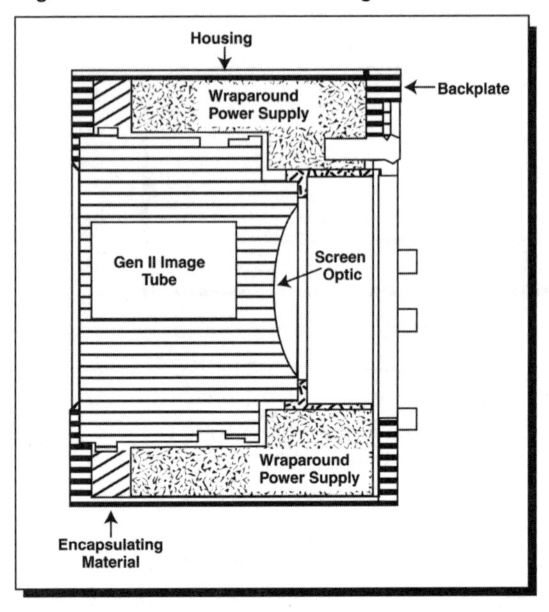

BACKGROUND AND OBJECTIVES

The customer required a rigorous environmental test which involves thermal excursions and a large range of relative humidity exposure to the viewing device as well as the image sensor module. Image sensor module reliability was defined at the customer's quality level as the ability of the device to maintain luminous gain over time. (Luminous gain is the light amplifying power of the image sensor module.)

The historical approach for improving the reliability of the product and image sensor module was to evaluate design and process changes while exposing these devices to the customer's environmental test. These evaluations were typically single factorial in nature and the quality characteristic measured was "number of cycles" until the device's "luminous" gain degraded past a certain value or "the total % of gain degradation." Prior to this effort, experiments used to improve the image sensor's performance against environmental factors measured the luminous gain change caused by the penetration of moisture into the module. Subsequent experimentation focused on measuring leakage current using the actual sensor product with marginal success. Finally, a coupon was developed to investigate the ideal function of the encapsulant in order to develop the encapsulating/insulating technology independent of the product.

At the engineering quality level, the quality characteristic used was interface leakage current measured at several voltage potentials applied between variably spaced electrodes on a sample coupon, not the image module itself. Figure 5 shows a representation of the coupon used to gather the data needed to generate the double-signal factor signal-to-noise ratio used to indicate the best levels of the controllable factors of the encapsulating/insulating engineered system. The coupon was designed on a ceramic substrate and consisted of a set of electrode lines deposited on its surface with spacings increasing in distance. The spacings represent, in a way, the noise factor of the distance between electrical elements on which the encapsulant would be placed. For the purposes of this parameter design effort, the distance between the electrodes becomes a second signal factor rather than a noise factor. If the space between electrodes acts as a resistor, then doubling the distance should double the resistance, and the voltage/current relationship at that distance should be linear.

Figure 5: Experimental Coupon

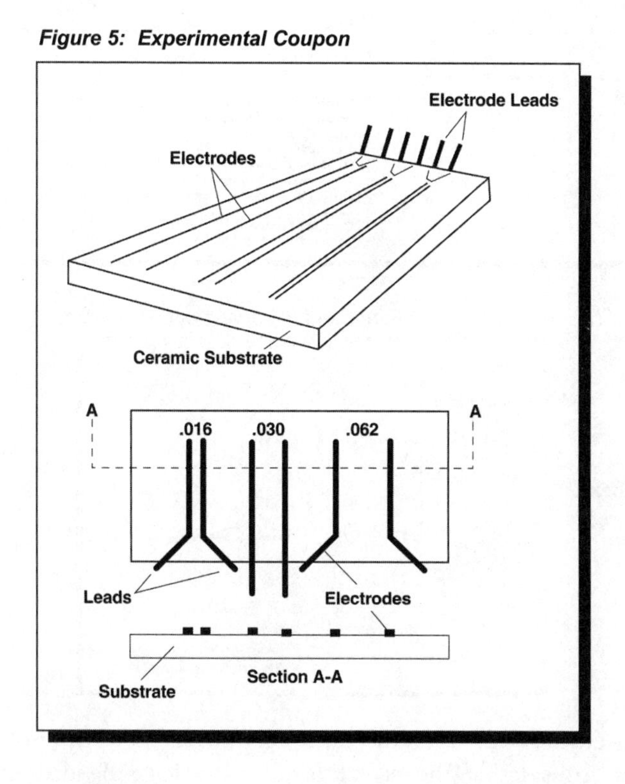

The objective of the study was to select the best levels of the system's controllable factors which bring the response (current leakage) as close to the ideal function as possible while it is exposed to environmental factors and time. The approach was to maximize the dual-signal factor signal-to-noise ratio.

Parameter Design Experimental Approach

Figure 6 presents the encapsulating/isolating engineered system. The signal factors are applied voltage and electrode spacing, both of which are really noise considering the product design. The actual electrical element spacing changes on the product as well as the voltage applied between them. The response is the resultant current flow between the elements. The ideal function for the system is presented in Figure 7. The primary noise factor is thermal/humidity exposure, performed cyclically over many days. The electrical data were collected before and after this cycling.

Figure 6: Encapsulant/Electrical Isolation Engineered System

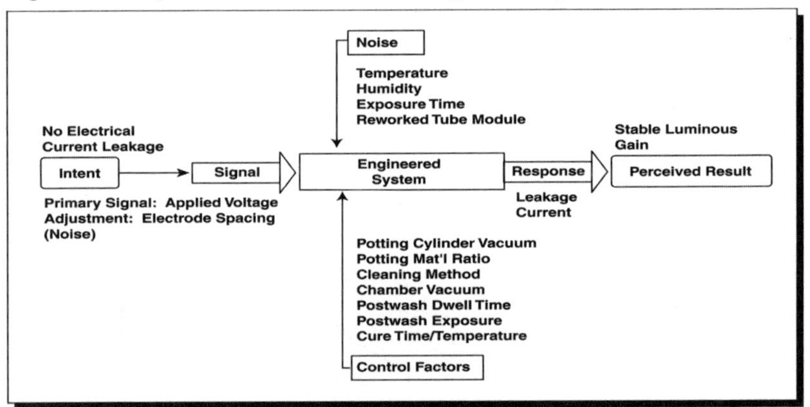

Figure 7: Ideal Function for Encapsulation System

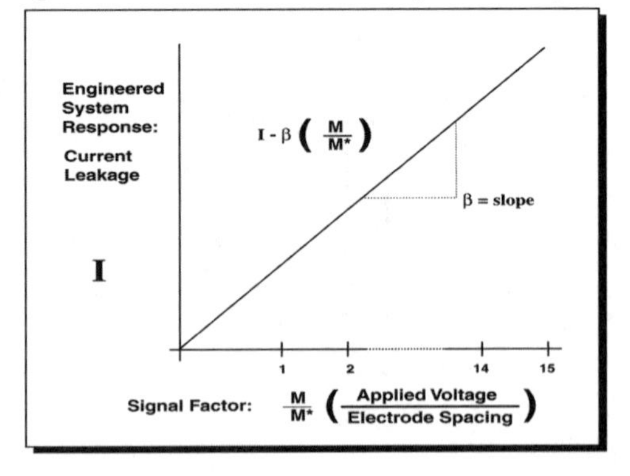

The data were analyzed using the zero-point proportional dynamic signal-to-noise ratio considering the signal factor as voltage divided by M*, the electrode spacing. The objective was to maximize the signal-to-noise ratio and minimize the slope, β. For the ideal function of voltage versus current, the slope is inversely proportional to resistance, and for this system, the objective is to maximize resistance in order to increase the electrical isolation properties of the encapsulant system.

Experimental Layout

Table 1 lists the factors and levels used in the L_{18} experimental layout. The signal factors are represented by outside orthogonal groups with column assignments shown in Figure 8.

Table 1: Experimental Factors and Levels

Signal Factors		
M	Voltage between Electrodes	Level 1: 2,000 Level 2: 4,000 Level 3: 6,000 Level 4: 8,000 Level 5: 10,000
M*	Spacing between Electrodes	Level 1: 0.016" Level 2: 0.030" Level 3: 0.062"
Control Factors		
A	Potting Cylinder Vacuum	Level 1: Yes Level 2: No
B	Potting Ratio	Level 1: Standard Level 2: Medium Level 3: High
C	Tube Cleaning Procedure	Level 1: A/P/DI Level 2: A Clean Level 3: A Clean 2
D	Potting Chamber Vacuum	Level 1: 600 mtorr Level 2: 2 torr Level 3: 10 torr
E	Postwash Delay	Level 1: 0 Level 2: 2 h Level 3: 4 h
F	Postwash Condition	Level 1: Room Level 2: Nitrogen Level 3: Humidity
G	Cure Time (hr) (Sliding Levels with Temp.)	Level 1: S– 6 5 4 Level 2: M– 8 7 6 Level 3: L–10 9 8
H	Cure Temperature (°C)	Level 1: 85 Level 2: 95 Level 3: 105
Noise Factors		
X	Humidity/Temperature	Level 1: Before Level 2: After
Y	Coupon Condition	Level 1: New Level 2: Reworked

Figure 8: Experimental Layout and Assignment of Factors

Primary Signal Factor									M1												…	M5											
Secondary Signal Factor									M*1				M*2				M*3				…	M*1				M*2				M*3			
Noise									X1		X2		X1		X2		X1		X2		…	X1		X2		X1		X2		X1		X2	
Control Factor Row	A	B	C	D	E	F	G	H	Y1	Y2	Y1	Y2	Y1	Y2	Y1	Y2	Y1	Y2	Y1	Y2		Y1	Y2	Y1	Y2	Y1	Y2	Y1	Y2	Y1	Y2	Y1	Y2
1	1	1	1	1	1	1	1	1																									
2	1	1	2	2	2	2	2	2																									
3	1	1	3	3	3	3	3	3																									
4	1	2	1	1	2	2	3	3																									
5	1	2	2	2	3	3	1	1																									
6	1	2	3	3	1	1	2	2																									
7	1	3	1	2	1	3	2	3																									
8	1	3	2	3	2	1	3	1																									
9	1	3	3	1	3	2	1	2																									
10	2	1	1	3	3	2	2	1																									
11	2	1	2	1	1	3	3	2																									
12	2	1	3	2	2	1	1	3																									
13	2	2	1	2	3	1	3	2																									
14	2	2	2	3	1	2	1	3																									
15	2	2	3	1	2	3	2	1																									
16	2	3	1	3	2	3	1	2																									
17	2	3	2	1	3	1	2	3																									
18	2	3	3	2	1	2	3	1													…												

Thirty-six coupons were fabricated, half of them being treated to represent one of the noise factors called "reworked." Each was encapsulated according to the orthogonal combinations using the process developed for sensor module fabrication. Once these were done, all the electrical measurements were obtained and then the coupons were exposed to repeated cycles of temperature and humidity. Each was then tested again for the electrical characteristics and the data collection sheets were completed (Figure 14).

Experimental Results and Analysis

The electrical current measurements, made on the 36 coupons for each voltage potential before and after thermal/humidity exposure, are too numerous to present in this report. The signal-to-noise ratio and β were calculated using a commercially available spreadsheet software (Excel) using the equations shown in Figure 9. The data layout of Figure 8 was rearranged as

Figure 9: Equation for Calculating the S/N Ratios for Each Experimental Orthogonal Row

$$\eta = 10 \log \frac{\frac{1}{r}(S_\beta - V_e)}{V_N}, \quad S_\beta = \frac{(L_1 + L_2 + L_3 + L_4)^2}{4r}, \quad S_{N\times\beta} = \frac{L_1^2 + L_2^2 + L_3^2 + L_4^2}{r} - S_\beta$$

$$S_T = \sum_{j=1}^{n} y_j^2, \quad S_e = S_T - S_\beta - S_{N\times\beta}, \quad V_e = \frac{S_e}{df_e}, \quad V_N = \frac{S_e + S_{N\times\beta}}{df_e + df_{N\times\beta}}$$

$$L_i = \left(\frac{M_1}{M_1^{*}}\right) y_{i,1} + \ldots + \left(\frac{M_5}{M_3^{*}}\right) y_{i,15} \quad i = 1, 2, 3, 4, \quad \beta = \frac{L_1 + L_2 + L_3 + L_4}{4r}$$

$$r = \left(\frac{M_1}{M_1^{*}}\right)^2 + \ldots + \left(\frac{M_5}{M_3^{*}}\right)^2, \quad y_j = \textit{data point}, \quad n = \textit{total number of data points}$$

shown in Figure 10 to facilitate the computations. A typical spreadsheet example of one of the orthogonal rows is presented in the Appendix.

This section is divided into two parts. The first presents analysis of the process and design baseline of the existing encapsulating system done in order to confirm sensitivity to the signal-to-noise ratio and slope calculation and to assess measurement resolution. The second presents the analysis of the L_{18} experiment.

Process Baseline

Figure 11 presents the ANOVA table used to calculate the signal-to-noise ratio using the equations of Figure 9.

Figure 10: Layout Used to Calculate the S/N Ratio and Slope β for Each Row of the Experiment

Compounded Noise Condition	Secondary Signal Factor (Electrode Spacing)	Primary Signal Factor (Voltage)				
		M1	M2	M3	M4	M5
N1	M*1	$y_{1,1}$	$y_{1,2}$	$y_{1,3}$	$y_{1,4}$	$y_{1,5}$
	M*2	$y_{1,6}$	$y_{1,7}$	$y_{1,8}$	$y_{1,9}$	$y_{1,10}$
	M*3	$y_{1,11}$	$y_{1,12}$	$y_{1,13}$	$y_{1,14}$	$y_{1,15}$
N2	M*1	$y_{2,1}$	$y_{2,2}$	$y_{2,3}$	$y_{2,4}$	$y_{2,5}$
	M*2	$y_{2,6}$	$y_{2,7}$	$y_{2,8}$	$y_{2,9}$	$y_{2,10}$
	M*3	$y_{2,11}$	$y_{2,12}$	$y_{2,13}$	$y_{2,14}$	$y_{2,15}$
N3	M*1	$y_{3,1}$	$y_{3,2}$	$y_{3,3}$	$y_{3,4}$	$y_{3,5}$
	M*2	$y_{3,6}$	$y_{3,7}$	$y_{3,8}$	$y_{3,9}$	$y_{3,10}$
	M*3	$y_{3,11}$	$y_{3,12}$	$y_{3,13}$	$y_{3,14}$	$y_{3,15}$
N4	M*1	$y_{4,1}$	$y_{4,2}$	$y_{4,3}$	$y_{4,4}$	$y_{4,5}$
	M*2	$y_{4,6}$	$y_{4,7}$	$y_{4,8}$	$y_{4,9}$	$y_{4,10}$
	M*3	$y_{4,11}$	$y_{4,12}$	$y_{4,13}$	$y_{4,14}$	$y_{4,15}$

Figure 11: ANOVA Table Used to Calculate the Signal-to-Noise Ratio

Source	df	S	V
β	1	S_β	V_β
$N \times \beta$	3	$S_{N \times \beta}$	$V_{N \times \beta}$
e	56	S_e	V_e
Total	60	S_T	V_T

There are 60 total data points used in the calculation for S_β, $S_{N \times \beta}$, S_e, and S_T.

Before the parameter design experiment began, the existing encapsulation system was baselined by producing coupons under standard conditions and making the measurements shown in Table 2. In addition to standard conditions, measurements were also taken at increased curing time to examine the impact of that factor.

Table 2: Baseline Measurements of the Standard Encapsulation System

Current (10^{-11}A) at 8- and 12-Hour Cures, Measured 2 Hours After Temperature/Humidity Exposure											
Electrode D	Cure Time	2 kV		4 kV		6 kV		8 kV		10 kV	
		Before	After	Before	After	Before	After	Before	After	Before	After
0.016"	8 hr	3	449	9	1085	19	1333	31	1701	52	3300
	12 hr	2	22	5	66	10	140	19	260	25	500
0.030"	8 hr	3	1075	7	1025	11	995	18	1325	25	2750
	12 hr	1	19	3	43	7	74	12	116	19	173
0.062"	8 hr	3	885	6	1675	9	2600	13	3420	18	4535
	12 hr	2	17	3	35	6	60	10	87	13	102

The signal-to-noise ratio and β for both the 8-hr and 12-hr cure times are –53.2 dB versus –48.9-dB and 3.01 versus 0.304, respectively, representing a 4.3 dB signal-to-noise improvement and a decrease in slope of 2.7.

Figure 12 is a graphical representation of the 8-hr and 12-hr curing time response to the dual-signal factor of voltage divided by electrode spacing. The slope of the 12-hr cure is much lower than the 8-hr cure and the influence of the noise factor, before and after thermal/humidity cycling, is greatly diminished, indicating a much more robust engineered encapsulation system.

Figure 13 presents the factor level average graphs of Table 2 indicating a significant influence due to curing time.

Figure 14 represents the experimental execution and data collection sheet used to record the data taken before and after thermal/humidity cycling.

Figure 12: Current vs. Signal for 8- vs. 12-Hour Cure Time

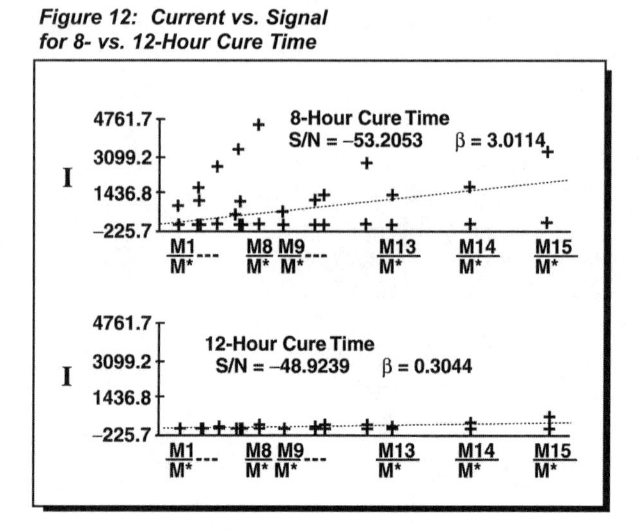

Figure 13: Factor Level Average Response Graphs for the Data in Table 2

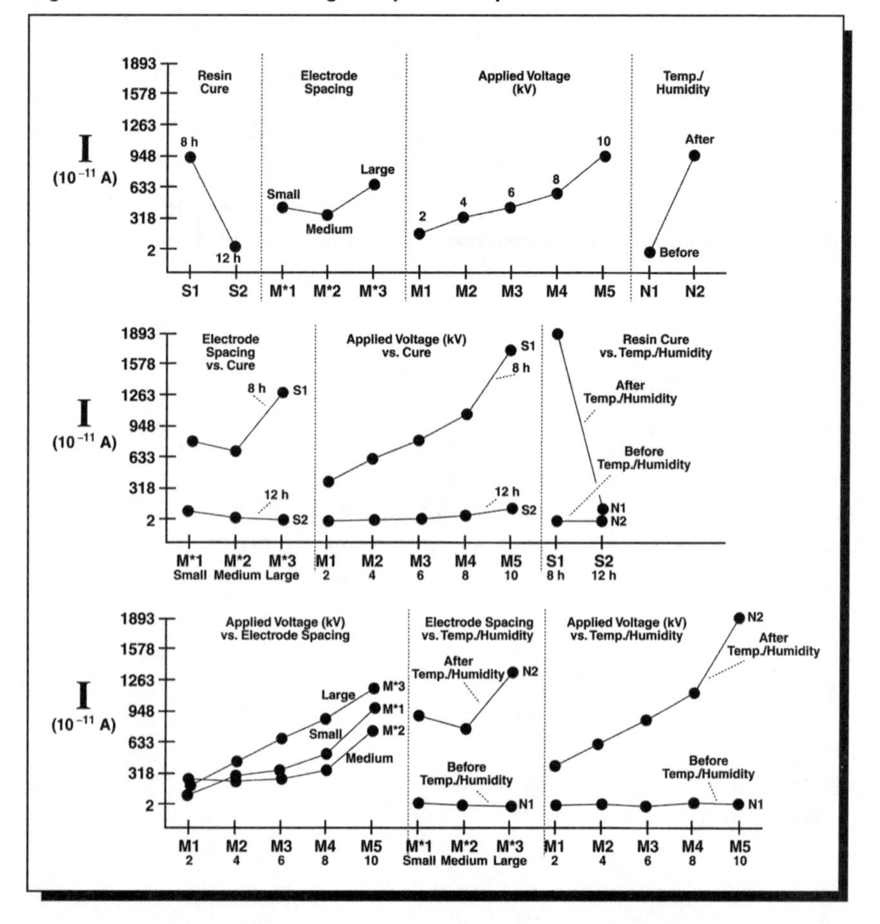

Figure 14: Typical Experimental Execution and Data Collection Sheet

Cell Name: L18#1
Experiment: [567 Potting Cure]
Engineer: [flora]
Char: [Voltage/Current]
Units: [Nanoamps]
Data Class: [Variable]
S/N Ratio: [ ]

Group 1 - [Orthogonal Array]
[L_{18}]

A: Potting Cyl. Vac. =	Yes
B: Potting Ratio =	Standard
C: Tube Clean =	A/P/DI
D: Potting Cham. Vac. =	600 mtorr
E: Postwash Delay =	0
F: Postwash Cond. =	Room
G: Cure Time (h) =	S– 6 5 4
H: Cure Temp. =	85

New Coupons: Current (10^{-11}A) Measured 2 Hours After Temp./Humidity Exposure											
Electrode D	Cure Time	2 kV		4 kV		6 kV		8 kV		10 kV	
		Before	After	Before	After	Before	After	Before	After	Before	After
Small											
Medium											
Large											

Used Coupons: Current (10^{-11}A) Measured 2 Hours After Temp./Humidity Exposure											
Electrode D	Cure Time	2 kV		4 kV		6 kV		8 kV		10 kV	
		Before	After	Before	After	Before	After	Before	After	Before	After
Small											
Medium											
Large											

η AND β ANOVA TABLES AND RESPONSE GRAPHS

Table 3 presents the η and β values which resulted from the L_{18} experiment calculated using the equations shown in Figure 9. Of particular importance in Table 3 is the large range in β values. It was the objective of the study to find factors which influence η and β, adjusting their levels to maximize η and minimize β.

Tables 4 and 5 present the ANOVA for η and β, respectively. The major contributor to η, the signal-to-noise ratio, is factor A, the potting cylinder vacuum. Factor A was also the major contributor to β, the slope.

The factor level average response graphs for η and β are presented as Figure 15. The best levels, focusing on minimizing β, are A1, B1, C3, D2, E3, F1, G3, H3.

Table 3: L_{18} η and β Results

L_{18} Row	η	β
1	–48.138	1.572
2	–47.445	0.875
3	46.799	0.399
4	–46.656	0.486
5	–47.802	1.084
6	–46.923	0.664
7	–46.998	0.624
8	–50.488	1.557
9	–47.619	0.911
10	–48.418	1.328
11	–47.734	0.818
12	–48.528	0.714
13	–49.395	1.538
14	–53.798	3.328
15	–53.259	4.481
16	–54.318	4.708
17	–47.579	0.601
18	–47.850	0.838

Table 4: ANOVA for η, Dual-Signal Factor S/N Ratio

Source	Df	S	V	F	S'	ρ
A. Cylinder Vacuum	1	26.9	26.9	39.4	26.2	25.65
B. Potting Ratio	2	10.3	5.1	7.5	8.9	8.73
C. Tube Cleaning	2	1.4	0.7	---	---	---
D. Chamber Vacuum	2	14.8	7.4	10.8	13.4	13.11
E. Postwash Delay	2	15.1	7.6	11.1	13.7	13.45
F. Wash Conditions	2	3.4	1.7	2.5	2.0	1.98
G. Cure Time	2	12.3	6.2	9.0	11.0	10.72
H. Cure Temperature	2	2.6	1.3	1.9	1.3	1.23
A x B. A, B Interaction	2	15.4	7.7	11.3	14.1	13.76
(e) Pooled Error	2	1.4	0.7	---	11.6	11.36
Total	17	102.2	6.0	---	---	100.00

Table 5: ANOVA for β, Slope of Current vs. Signal Response

Source	Df	S	V	F	S'	ρ
A. Cylinder Vacuum	1	5.760	5.760	22.796	5.507	18.67
B. Potting Ratio	2	2.916	1.458	5.770	2.410	8.17
C. Tube Cleaning	2	0.505	0.253	---	---	---
D. Chamber Vacuum	2	3.319	1.660	6.569	2.814	9.54
E. Postwash Delay	2	4.286	2.143	8.482	3.780	12.82
F. Wash Conditions	2	2.781	1.391	5.504	2.276	7.72
G. Cure Time	2	3.738	1.869	7.397	3.232	10.96
H. Cure Temperature	2	1.960	0.980	3.879	1.455	4.93
A x B. A, B Interaction	2	4.228	2.114	8.368	3.723	12.62
(e) Pooled Error	2	0.505	0.253	---	4.295	14.56
Total	17	29.493	1.735	---	---	100.00

Figure 15: Factor Level Average Response Graphs for η and β (Circled Values Are Standard)

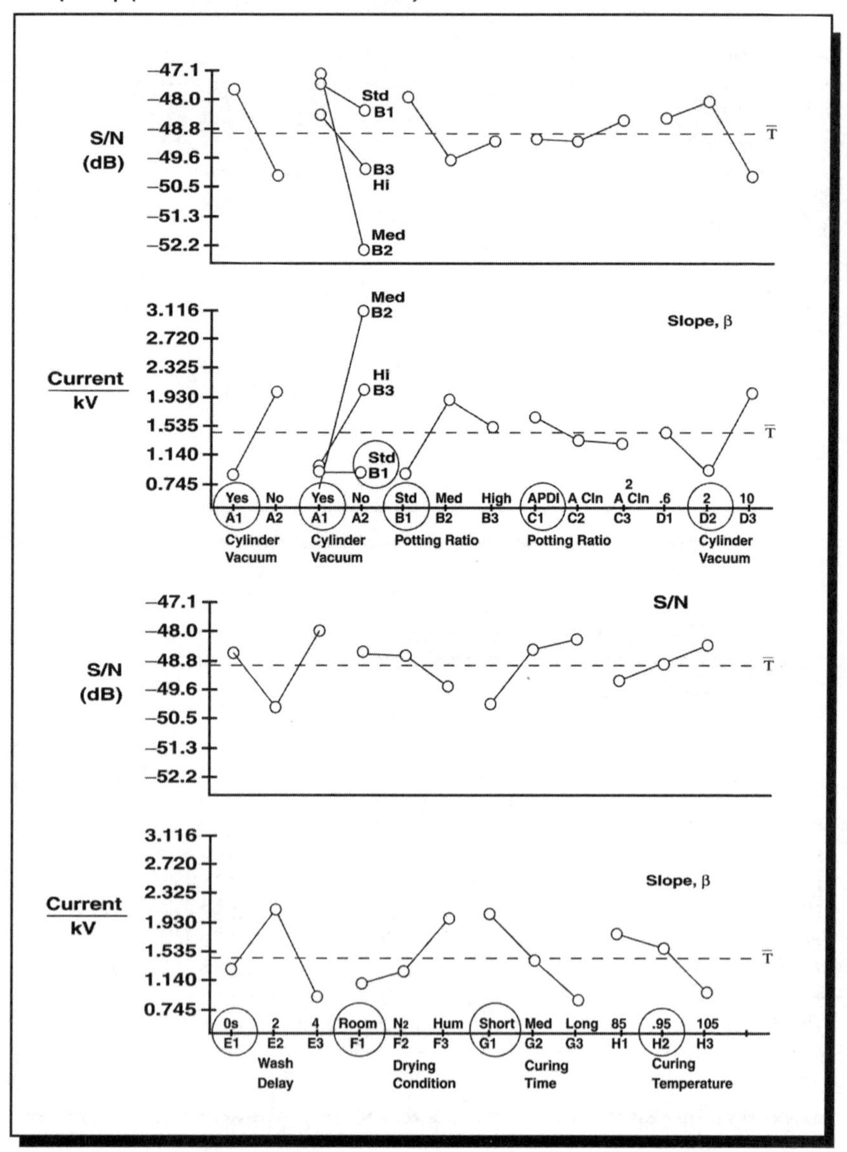

CONFIRMATION

Several confirmation conditions were prepared in order to (1) confirm the analysis with the coupons, (2) obtain the standard conditions for comparison, and (3) test the condition where the temperature is limited to 95°C due to product subassembly limitations and a time extension greater than the eight hours of the experiment.

Table 6 presents these conditions along with the average estimates of the process for η and β and the values obtained during confirmation runs. Figure 16 presents some of the information in Table 6 to graphically compare the improvements for the "best" conditions obtained compared to the "standard" process conditions.

Table 6: Confirmation Estimates and Values

Condition	Process Average Estimate		Confirmed Value	
	η (dB)	β	η (dB)	β
Standard A1 B1 C1 D2 E1 F1 G1 H2	--- 50.5 to --- 43.8	0 to 2.61 8	--- 50.1 1 ---	1.37 3 --- ---
Optimum A1 B1 C3 D2 E3 F1 G3 H3	--- 47.4 to --- 40.7	0 to 0.61 4	--- 48.5 1 ---	1.01 0 --- ---
95° C, 9 Hrs A1 B1 C3 D2 E3 F1 G3+ H2	--- 47.9 to --- 41.2	0 to 1.17 4	--- 52.7 0 ---	1.38 5 --- ---
95° C, 12 Hrs A1 B1 C3 D2 E3 F1 G3++ H2	--- --- ---	--- --- ---	--- 46.3 5	0.27 9 ---

Figure 16: "Best" vs. Standard Ideal Function Graphs

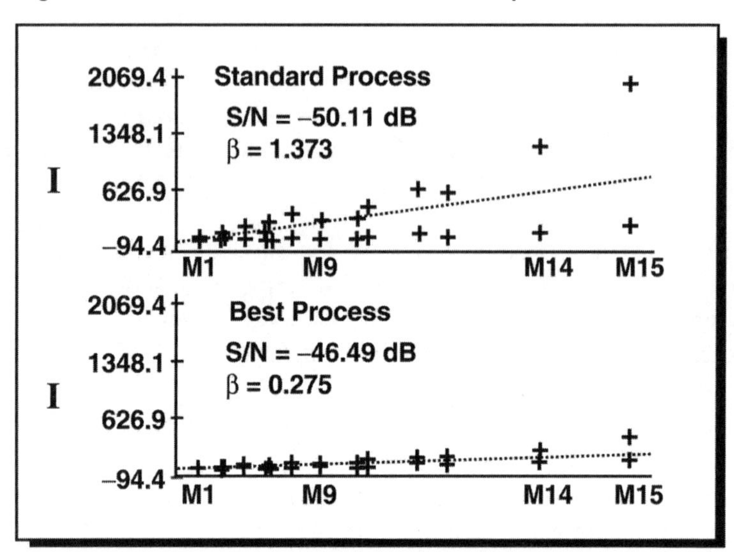

CONCLUSION

As a result of the parameter design effort focused on the electrical encapsulation engineered system, the dual-signal factor signal-to-noise ratio increased 3.8 dB while decreasing the slope, β, 80%, from 1.373 to 0.275.

Of far more importance is the improvement in encapsulation technology since the parameter design effort was performed on coupons and focused on improving the function of the system independent of product application. Confirmation at the customer's quality level is underway to measure the improvement in duration of luminance gain.

This case study is contributed by Lapthe Flora and Ron Ward of ITT Night Vision, Roanoke, VA, and Tim Reed of ITT Defense and Electronics, McLean, VA.

CHAPTER FIFTEEN

Temperature-Rising Problem for a Printer Light-Generating System

MINOLTA CO. – JAPAN

EXECUTIVE SUMMARY

BACKGROUND

During the development stage of a printer, it was found that the temperature rise in the light source area was much higher than expected. Since the countermeasures that are available would result in a cost increase, it was decided to apply robust design to reduce the temperature. Normally, trying to lower the temperature requires temperature measurement. Such an approach is not recommended because of two reasons. First, the environmental temperature must be controlled during experimentation. Second, the selection of material must consider all three aspects of heat transfer, i.e., conduction, radiation and convection. Hence, this process is quite time consuming.

To reduce the development time, it was decided to consider the functionality of an airflow (air speed) generation by the cooling fan.

PROCESS

Instead of following the common design–build–test method, the technically more efficient robust design methodology was used and the functionality of the engineering system was studied. Eight control factors were tested by 18 design configurations under two noise levels, with and without obstacle. Motor voltage into the fan was considered the input signal and air speed was measured as the output response.

BENEFITS

- The developmental time was reduced significantly with the help of the robust design approach. It took only four hours to complete the study including the fabrication of the components from cardboard used to build the test pieces.
- A 5.83-dB gain in the S/N ratio was achieved.
- An 11.1-dB improvement in air speed generation was achieved. This is equivalent to an increase in efficiency of 360%.

- As a result of improvement in air speed generation, the temperature in the area of the light source was reduced from 115°C to less than100°C.
- The result was confirmed.

CASE STUDY

INTRODUCTION

During the development stage of a printer, it was noticed that the temperature rise in the light source area was much higher than expected. To solve this problem, there were some possible countermeasures, such as upgrading the resin material to retard flammability or adding a certain heat-resisting parts. Since these countermeasures would result in a cost increase, it was decided to reduce the temperature.

Normally, trying to lower the temperature requires temperature measurement. Such an approach is not recommended because of two reasons. First, the environmental temperature must be controlled during experimentation. Second, the selection of material must consider all three aspects of heat transfer, i.e., conduction, radiation and convection. Hence, this process is quite time consuming.

To reduce the development time, it was decided to consider the functionality of airflow (speed) generation by the cooling fan.

Description of the System and the Ideal Situation

Figure 1 describes the system. Primarily, the main system consists of two subsystems: a light-generating system and a cooling system. Since the light-generating system is a standard purchased item, it was decided to improve the cooling system.

The objective function is described as "airflow generated by the fan to remove the air surrounding the heat source."

Therefore, the ideal situation is "the air speed surrounding the heat source is proportional to the fan RPM with high efficiency (sensitivity)."

But to measure the fan RPM, we had to purchase an RPM meter. Therefore, it was decided simply to vary motor voltage, since motor voltage is likely proportional to RPM. Therefore, the ideal function was redefined as "the air speed surrounding the heat source is proportional to motor voltage with a high efficiency (sensitivity)." This ideal function is shown in Figure 2.

Figure 1: The System *Figure 2: Ideal Function*

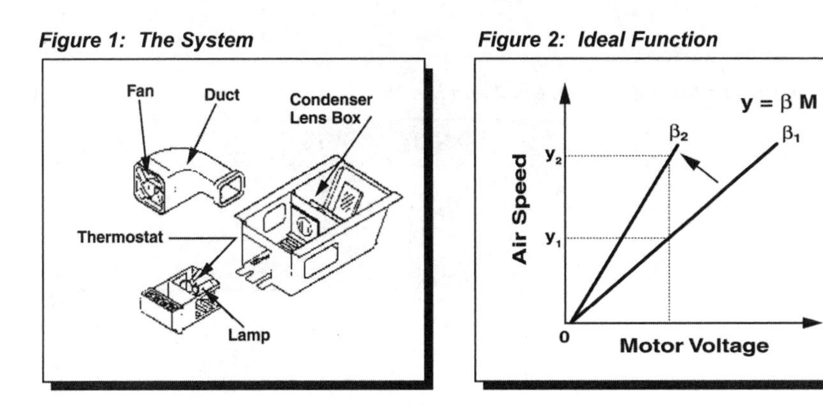

Description of Factors

Signal Factor:

M = Motor Voltage

M_1 = 5 Volts, M_2 = 15 Volts, M_3 = 25 Volts

Noise Factor:

N = Obstacle

N_1 = No, N_2 = Yes

Control Factors

Factors	Level 1	Level 2	Level 3
A. Damper	No*	Yes	---
B. Distance between Suction and Outer Case	20*	40	60
C. Distance between Suction and Heat Source	110*	60	40
D. Opening Height	30*	15	0
E. Duct Height	30*	15	0
F. Size of Hole above Heat Source	Large*	Medium	Small
G. Size of Hole below Heat Source	No*	Medium	Large
H. Distance between Heat Source and Duct	60*	50	40

** Existing condition*

Layout and Results:

Control factors were assigned to an L_{18} array. Table 1 shows results from runs No. 1, No. 2 and No. 18. Table 2 shows S/N ratios and sensitivities for each run of L_{18}. Figures 3 and 4 show the response graphs of S/N ratios and sensitivities, respectively.

Table 1: Results for Runs No. 1, No. 2 and No. 18

No.	Air Speed (m/s x 100)					
	M_1 = 5 V		M_2 = 15 V		M_3 = 25 V	
	N_1	N_2	N_1	N_2	N_1	N_2
1	12	9	31	26	44	41
2	18	15	28	23	44	32
---	---	---	---	---	---	---
---	---	---	---	---	---	---
18	28	19	76	57	106	71

Table 2: S/N Ratio and Sensitivity

	A	B	C	D	E	F	G	H	S/N	Sensitivity
	1	2	3	4	5	6	7	8		
1	1	1	1	1	1	1	1	1	−4.17	−35.08
2	1	1	2	2	2	2	2	2	−12.77	−35.86
3	1	1	3	3	3	3	3	3	−5.99	−23.94
4	1	2	1	1	2	2	3	3	1.76	−26.29
5	1	2	2	2	3	3	1	1	−4.81	−26.36
6	1	2	3	3	1	1	2	2	−5.35	−26.74
7	1	3	1	2	1	3	2	3	−15.93	−35.41
8	1	3	2	3	2	1	3	1	−14.45	−30.67
9	1	3	3	1	3	2	1	2	−5.35	−26.15
10	2	1	1	3	3	2	2	1	−8.82	−26.58
11	2	1	2	1	1	3	3	2	−11.40	−37.24
12	2	1	3	2	2	1	1	3	−1.08	−23.41
13	2	2	1	2	3	1	3	2	−5.57	−27.06
14	2	2	2	3	1	2	1	3	−4.92	−23.97
15	2	2	3	1	2	3	2	1	−8.00	−33.99
16	2	3	1	3	2	3	1	2	−9.13	−24.54
17	2	3	2	1	3	1	2	3	−4.89	−26.25
18	2	3	3	2	1	2	3	1	−11.99	−28.41

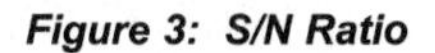
Figure 3: S/N Ratio

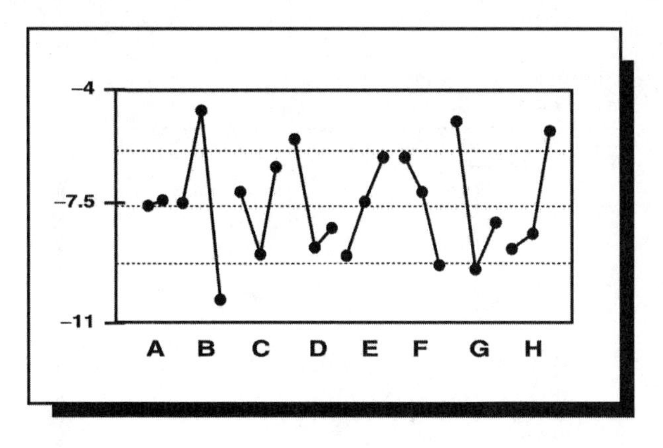

Figure 4: Sensitivity Response

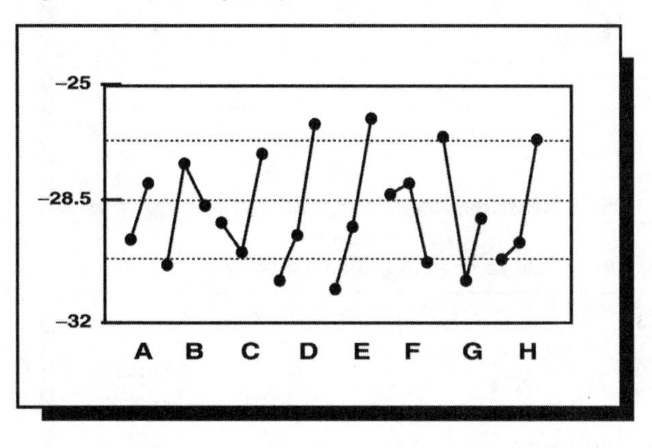

CONFIRMATION

Table 3 shows the results of prediction and confirmation.

Figure 5 shows the air speed of initial and optimum conditions. Figure 6 shows the temperature before and after the experiment.

Table 3: Prediction and Confirmation

Design	S/N Ratio		Sensitivity	
	Prediction	Confirmation	Prediction	Confirmation
Current	–4.3	–4.17	–32.4	–35.0
Optimum	2.23	1.66	–23.7	–24.0
Gain	6.53	5.83	8.7	11.1

Figure 5: Comparison of Initial and Optimum Conditions

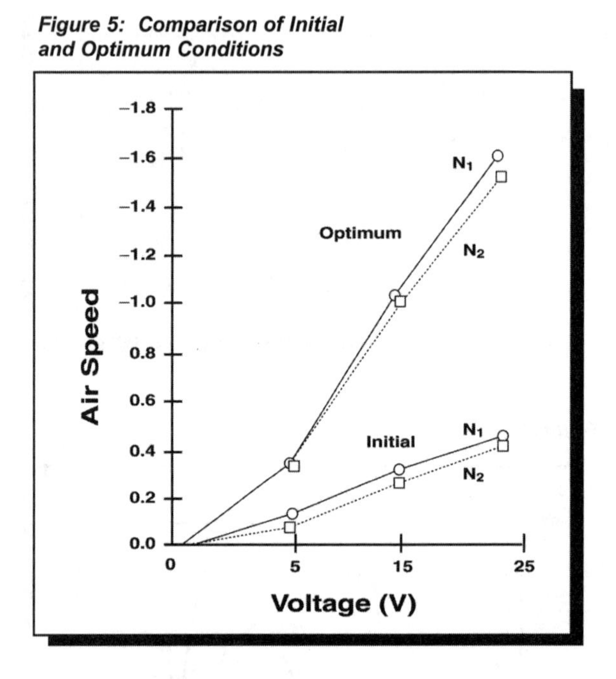

Figure 6: Temperature Before and After

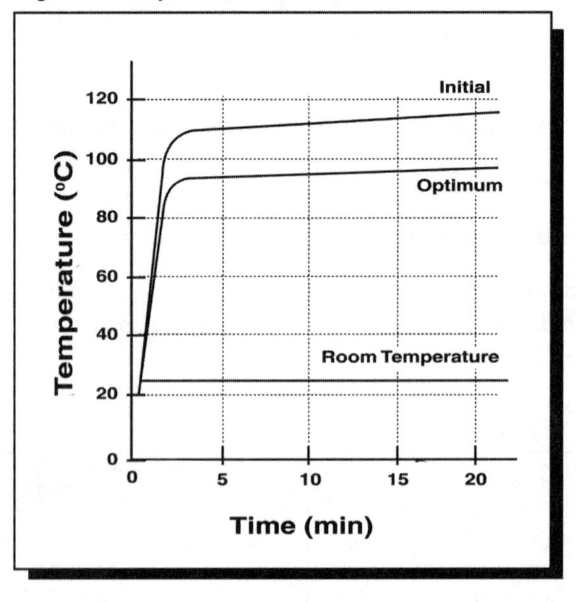

Conclusion

In order to reduce the development cycle time, cardboard was used to fabricate the cooling system since the airflow generation function will not be affected by material type. Further, those hand-assembled units probably enhanced the noise effect from manufacturing variability.

As a result, it took only four hours to complete the study including the fabrication of the components.

This case study is contributed by Hideki Kusano, Hajime Ootuki, Keiichiro Bungo, Shuichi Saito and Toshio Takaki of Minolta Company, Japan.

CHAPTER SIXTEEN 

Technology Development of the Drain Electrode Process for a Power MOSFET

NISSAN MOTOR CO., LTD., AND
NISSAN TOCHIGI INSTITUTE OF AUTOMOTIVE TECHNOLOGY – JAPAN

EXECUTIVE SUMMARY

BACKGROUND

A power MOSFET has been used in automotive electronics as a reliable switching device. An increase of electronics systems requires the power MOSFET with low on-resistance to reduce power dissipation. In general, on-resistance of the power MOSFET is reduced by miniaturization of unit cells. In this paper, however, it is shown that low on-resistance can be achieved by the improvement of the drain electrode process. Taguchi Methods™ have been applied to improve the drain electrode process. This result also can be applied to devices with similar structures.

PROCESS

Nine process parameters from the drain electrode were studied to optimize the conductivity. Eighteen combinations of nine parameters were tested for robustness. For each of 18 runs, the input currents were varied at 0.5, 1.0 and 1.5 Amperes, and voltage was measured under two compounded noise conditions. Since conductivity must be assured under various customers' usage environments including aging, two noise conditions were derived from compounding two noise factors, "ambient temperature" and "cycling."

BENEFITS

- Because the robust drain electrode was achieved, the annealing process was eliminated. (The annealing process was added as a follow-up solution for a poor drain electrode process. It is a typical engineering practice to add compensation to fix problems.)
- Because the annealing is eliminated, needless to say, all defects from annealing were eliminated (defects such as peeling and cracking from annealing).

- Because there is no need for annealing, the substrate can be thinned further (it does not have to be as strong as when annealing was practiced), which leads to a reduction of resistance that is due to the substrate.
- A total reduction of the on-resistance (substrate and drain electrode) of 25% was achieved.
- Since a 25% reduction in on-resistance over the life of product was achieved, the chip size can be reduced by 25% with the same power dissipation.
- The improvement provides a remarkable 25% cost reduction in unit manufacturing cost of power MOSFET.
- Moreover, this experiment made us realize that conventional engineering activities do not optimize for robustness and we are missing numerous great opportunities for cost reduction.

CASE STUDY

INTRODUCTION

A power MOSFET has been used in automotive electronics as a reliable switching device. An increase of electronics systems requires the power MOSFET with low on-resistance to reduce power dissipation. In general, on-resistance of the power MOSFET is reduced by miniaturization of unit cells. In this paper, however, it is shown that low on-resistance can be achieved by the improvement of the drain electrode process. Taguchi Methods have been applied to improve the drain electrode process. This result also can be applied to devices with similar structures.

EXPERIMENT

Background

In the vertical power MOSFET, current flows from a drain to source electrode as shown in Figure 1.

Figure 1: Current Flow of a Power MOSFET

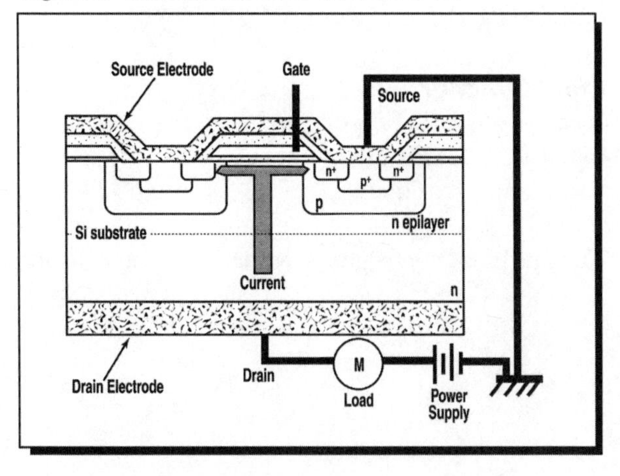

On-resistance consists of several components (R_1, R_2, etc.) as shown in Figure 2. The drain resistance, composed of the drain contact resistance R_1 and the

Figure 2: Components of the On-Resistance of a Power MOSFET

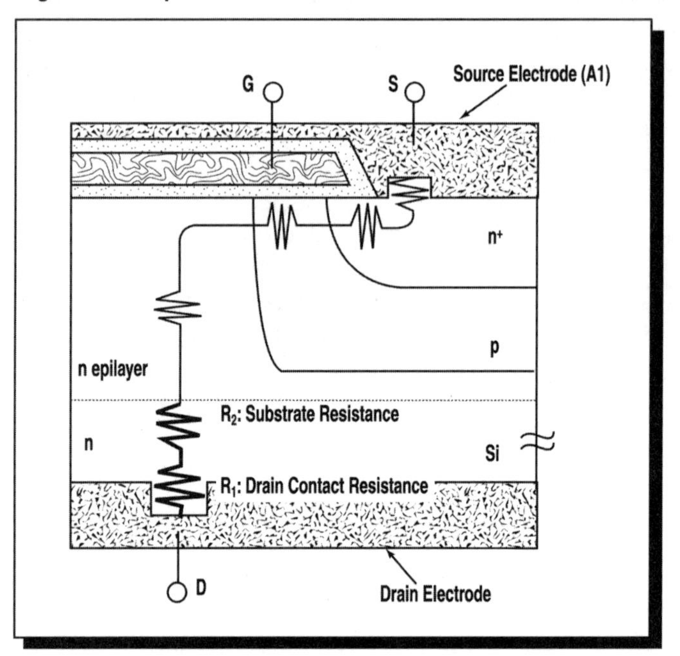

substrate resistance R_2, contributes substantially to the on-resistance of the power MOSFET. It is well known that the drain contact resistance R_1 is reduced by increasing the impurity concentration of the silicon substrate (by impurity doping to the silicon substrate and subsequent high temperature annealing), and the substrate resistance R_2 is reduced by thinning the silicon substrate. However, this process tends to cause warps or cracks of the silicon substrate due to mechanical stress during annealing in a fabrication process. Therefore, we investigated a new drain electrode process to reduce the drain contact resistance R_1 without impurity doping and high temperature annealing. This process allows thinning of the silicon substrate and reduces on-resistance of the power MOSFET.

Ideal Function

Ideally, voltage is proportional to current. This ideal function is shown in Figure 3. The purpose of this experiment is to maximize the S/N ratio and to minimize sensitivity.

The ideal function is given by the following equation:

$$y = \beta M$$

where the signal M is current, response y is voltage, and β is slope (i.e., resistance).

Figure 3: Ideal Function

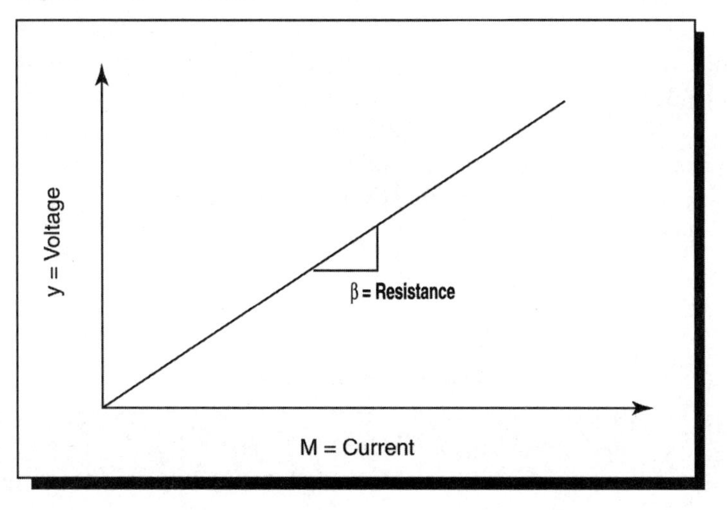

Figure 4 shows the measurement setup of this experiment. The current flows from the Cu stem to the Al electrode and the voltage drop between the Cu stem and the Al electrode are measured. The measured resistance includes both R_1 and R_2; however, the substrate resistance R_2 is negligible due to the low impurity concentration of the silicon substrate.

Figure 4: Measurement Steup

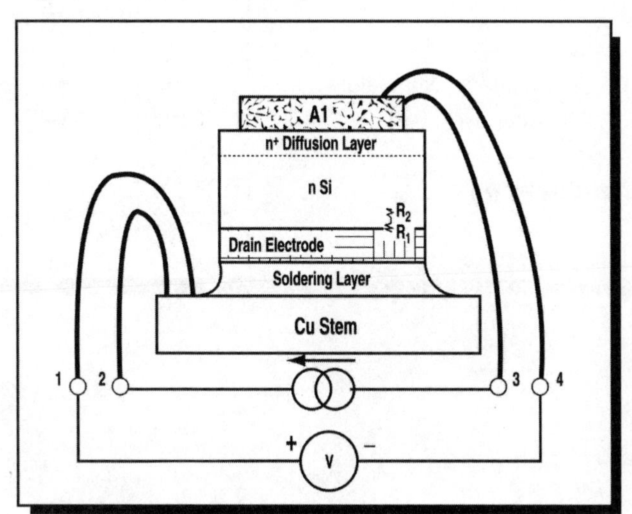

Control Factors, Noise Factors and Signal Factor

The control factors (A to H), noise factors (I, J), and signal factor (M) of this study are summarized in Table 1. Nine process parameters that may affect the drain contact resistance R_1 were selected for control factors. Atmosphere temperature and thermal cycle test were selected for noise factors. A compounded noise approach was utilized to reduce the number of tests. The compounded noise factors and summary of data (test No. 1 of L_{18}) are shown in Table 2. The S/N ratio and sensitivity were calculated for each of the 18 tests from L_{18}. The summary of the S/N ratio and sensitivity is shown in Table 3 along with the L_{18} matrix. Control factor effects are shown in Table 4 and Figure 5.

Table 1: Control, Noise and Signal Factors and Levels

Factors		Levels	1	2	3
Control Factors	A	Drain Electrode Materials	A*	B	---
	B	Temperature 1 (°C)	20	150	350*
	B'	Temperature 2	Low	Medium	High*
	C	Time 1 (min)	0*	1	5
	D	Time 2 (s)	0	20*	40
	E	Time 3 (min)	0*	1	5
	F	Chemical Treatment	None	A	B*
	G	Time 4 (min)	0	10*	20
	H	Temperature 3	Low	Low (Dummy)	High*
Noise Factors	I	Atmosphere Temperature	Low	High	----
	J	Thermal Cycle Test	Initial	After	----
Signal Factor	M	Current (A)	0.5	1.0	1.5

**Present.*

Table 2: Compounded Noise Factors and Summary of Data (Test No. 1) (mV)

Compounded Noise Factors \ Signal Factor (A)			M_1 0.5	M_2 1.0	M_3 1.5
High Resistance	I1 J1	(Low) (Initial)	428	523	597
Low Resistance	I2 J2	(High) (After)	276	356	418

Table 3: Summary of Test Results

Test No.	A	B	B'	C	D	E	F	G	H	S/N Ratio	Sensitivity (dB)
1	1	1	1	1	1	1	1	1	1	8.48	51.81
2	1	1	1	2	2	2	2	2	1'	21.07	29.97
3	1	1	1	3	3	3	3	3	3	20.47	28.46
4	1	2	2	1	1	2	2	3	3	13.37	48.67
5	1	2	2	2	2	3	3	1	1	25.54	24.50
6	1	2	2	3	3	1	1	2	1'	6.91	42.28
7	1	3	3	1	2	1	3	2	3	2.53	38.97
8	1	3	3	2	3	2	1	3	1	3.51	39.03
9	1	3	3	3	1	3	2	1	1'	14.15	30.10
10	2	1	2	1	3	3	2	2	1	10.75	59.80
11	2	1	2	2	1	1	3	3	1'	10.06	61.32
12	2	1	2	3	2	2	1	1	3	19.07	52.09
13	2	2	3	1	2	3	1	3	1'	7.14	50.01
14	2	2	3	2	3	1	2	1	3	6.53	48.14
15	2	2	3	3	1	2	3	2	1	9.20	55.76
16	2	3	1	1	3	2	3	1	1'	9.59	44.96
17	2	3	1	2	1	3	1	2	3	8.89	59.58
18	2	3	1	3	2	1	2	3	1	6.17	48.44

Table 4: Control Factor Effects

Control Factors	S/N Ratio			Sensitivity (dB)		
	1	2	3	1	2	3
A	12.89	9.71	----	37.09	53.34	----
B	13.46	12.26	8.19	46.88	43.89	44.87
B'	12.92	12.72	8.26	43.21	47.94	44.50
C	8.64	12.60	12.66	49.03	43.76	42.86
D	10.69	13.59	9.63	51.21	40.66	43.78
E	6.78	12.64	14.49	48.49	45.08	42.08
F	9.00	12.01	12.90	49.13	44.19	42.33
G	13.89	9.89	10.12	41.93	47.73	45.99
H	11.05	----	11.81	44.84	----	45.98
Overall Mean Value	11.30			45.22		

Figure 5: Control Factor Effects

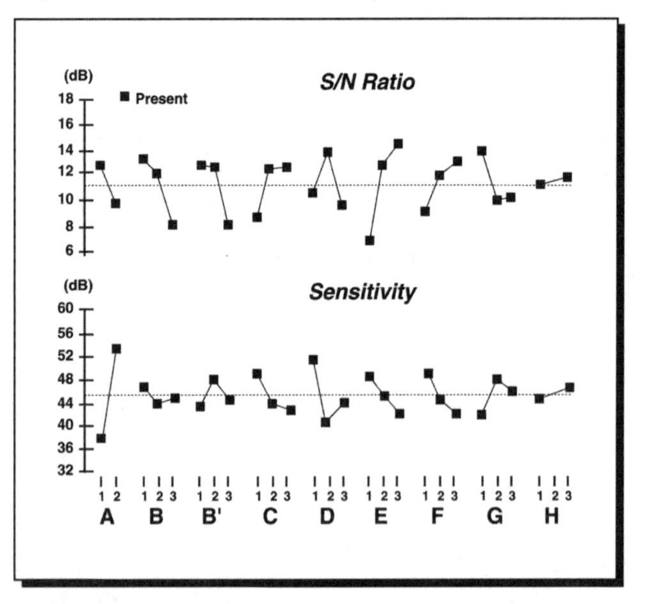

Optimum Configuration and Prediction of Effects

Both a high S/N ratio and a low sensitivity were chosen as the optimum configuration. The optimum configuration of the drain electrode process is as follows:

A1-B1-B'1-C3-D2-E3-F3-G1-H1

The S/N ratio of the optimum configuration (A1-B1-B'1-C3-D2-E3-F3-G1-H1) was predicted using five dominant factors (A, B, B', C, E) as follows:

$$\eta_{opt} = (\overline{A}_1 - \overline{T}) + (\overline{B}_1 - \overline{T}) + (\overline{B'}_1 - \overline{T}) + (\overline{C}_3 - \overline{T}) + (\overline{E}_3 - \overline{T}) + \overline{T}$$
$$= 21.22 \text{ dB}$$

The S/N ratio of the original configuration (A1-B3-B'3-C1-D2-E1-F3-G2-H3) was predicted using the same five factors (A, B, B', C, E) as follows:

$$\eta_{original} = (\overline{A}_1 - \overline{T}) + (\overline{B}_3 - \overline{T}) + (\overline{B'}_3 - \overline{T}) + (\overline{C}_1 - \overline{T}) + (\overline{E}_1 - \overline{T}) + \overline{T}$$
$$= -0.45 \text{ dB}$$

The sensitivity of the optimum configuration (Al-B1-B'l-C3-D2-E3-F3-G1-H1) was predicted using six dominant factors (A, C, D, E, F, G) as follows;

$$S_{opt} = (A1-T) + (C3-T) + (D2-T) + (E3-T) + (F3-T) + (G1-T) + T$$
$$= 20.87 \text{ dB}$$

The sensitivity of the original configuration (A1-B3-B'3-C1-D2-E1-F3-G2-H3) was predicted using the same six factors (A, C, D, E, F, G) as follows;

$$S_{original} = (A1-T) + (C1-T) + (D2-T) + (E1-T) + (F3-T) + (G2-T) + T$$
$$= 39.25 \text{ dB}$$

From these results, the predicted gain in the S/N ratio is 21.67 dB and the predicted gain in sensitivity is 18.38 dB.

Results of the Confirmation Test

The results of the confirmation test are summarized in Table 5. The actual gain in the S/N ratio from actual confirmation test runs is 19.61 dB and the actual gain in sensitivity is 17.18 dB. The results of the confirmation test are shown in Figure 6. The drain contact resistance R_1 becomes one-eighth of the present one in the optimum configuration. In addition, the drain contact resistance R_1 is not changed by the atmosphere temperature test and thermal cycle test. This means the characteristics of the power MOSFET are much more robust than the present ones.

Table 5: Results of Confirmation Test

	S/N Ratio		Sensitivity	
	Predicted	Actual	Predicted	Actual
Optimum	21.22	22.09	20.87	23.10
Present	−0.45	2.48	39.25	40.28
Gain	21.67	19.61	−18.38	−17.18

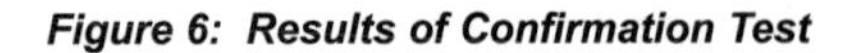
Figure 6: Results of Confirmation Test

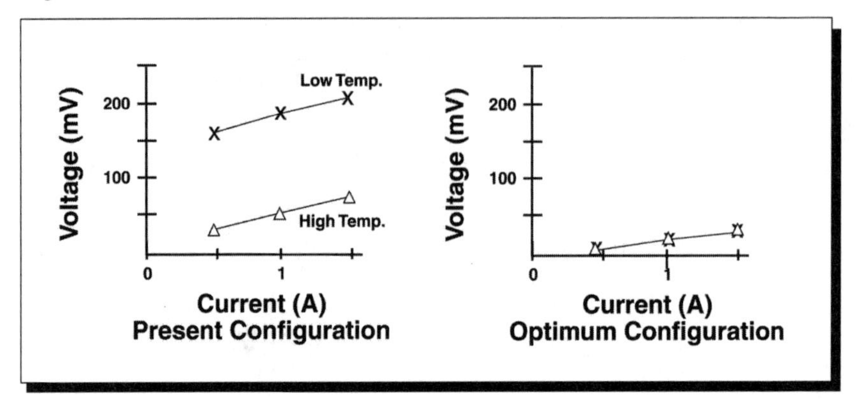

CONCLUSION

Taguchi Methods have been applied to improve the drain electrode process for the power MOSFET. By this experiment, a new drain electrode process to reduce drain contact resistance R_1 was established. Therefore, the silicon substrate can be thinned to reduce the substrate resistance R_2, since this process does not include annealing which causes warps and/or cracks of the silicon substrate by mechanical stress. The process is also able to reduce the device process steps. The total reduction of the on-resistance (R_1 and R_2) comes to approximately 25% of the present device. This means, compared to the present device with equal power dissipation, shrinking (25%) the chip size of the device is possible. Improving productivity by shrinking the chip size of the device and reducing the device process steps can lead to approximately a 25% cost reduction.

This case study is contributed by Koji Manabe, Kenji Yoa and Shigeo Hoshino of Nissan Motor Co., Ltd.., Japan, and Akio of Nissan Tochigi Institute of Automotive Technology, Japan. The authors would like to thank Teruyoshi Mihara and members of Nissan Electronics and Information Systems Research Laboratory for their support and discussions, and members of Nissan Reliability Engineering Center for their support and suggestions.

CHAPTER SEVENTEEN

Optimization of Bean Sprouting

SAMPO KAGAKU KOGYO CO. – JAPAN

EXECUTIVE SUMMARY

BACKGROUND

Plant-growing conditions have been traditionally determined by quick growth and abundant harvest, and bean sprouts are not an exception.

Bean sprouts are cultivated under an inhibitory condition called shaded cultivation without exposure to sunlight. The bean sprouts produced this way tend to break down and decay quickly as opposed to naturally grown bean sprouts. The faster the growth, the faster the decay, and the slower the growth, the smaller the product. These two characteristics contradict each other. It is therefore important to determine the optimum process condition to resolve this conflict.

PROCESS

In this study the basic function for bean sprout growth was determined. A transformation of this function was used for optimization of biological growth. Time was taken as the input signal and weight growth was the response. Seven control factors were tested using an orthogonal array L_{18} under at two levels of noise conditions (desiccators adjusted at different levels of humidity).

BENEFITS

- Increased productivity by 43%.
- Before this experiment, it took seven days to sprout. The optimum design reduced it to four days.
- It has developed a revolutionary way to evaluate biological growth function.
- A 2.2-dB gain in the S/N ratio was achieved.

CASE STUDY

INTRODUCTION

Bean sprouts are cultivated under an inhibitory condition called shaded cultivation without exposure to sunlight. The bean sprouts produced this way tend to break down and decay quickly as opposed to naturally grown bean sprouts. The faster the growth, the faster the decay, and the slower the growth, the smaller the product. These two characteristics contradict each other. It is therefore important to determine the optimum process condition to resolve this conflict.

GENERIC FUNCTION

Bean sprouts grow by absorbing moisture. The generic function of the growth was determined by

$$Y_s = Y_0 \, e^{\beta_T} \qquad (1)$$

where

Y_0 = initial weight

Y_s = the weight after T days

Equation (1) can be modified as

$$\ln (Y_s / Y_0) = \beta_T \qquad (2)$$

Letting the natural logarithm of (Y_s/Y_o) be y,

$$y = \beta_T \qquad (3)$$

Equation (3) is the ideal function. The dynamic S/N ratio based on this ideal function was used to evaluate the bean sprouting function.

Table 1 shows a list of control factors and levels for this optimization experiment.

Table 1: Control Factors

	Control Factors	Level 1	Level 2	Level 3
A	Type of Bean	A1	A2	
B	Indoor Temperature	B1 = 18°C	B2 = 24°C	B3 = 30°C
C	Number of Ethylene Gas Baths	C1 = Once (9 am)	C2 = Twice (9 am,12 pm)	C3 = ThreeTimes (9 am,12 pm,5 pm)
D	Concentration of Ethylene Gas	D1 = 10 (mL/1)	D2 = 20 (mL/1)	D3 = 30 (mL/1)
E	Water Spraying	E1 = Once (9 am)	E2 = Twice (9 am, 12 pm)	E3 = ThreeTimes (9 am,12 pm,5 pm)
F	Number of Water Sprayings	F1 = Once	F2 = Twice	F3 = Three Times
G	Concentration of Minerals in Water	G1 = 0%	G2 = 0.1%	G3 = 1.0%

Factors and Levels

Noise Factor

Since it is difficult to control the temperature and humidity of the cultivating room, the following noise factor was used:

N_1: Desiccators are adjusted at 60% humidity
N_2: Desiccators are adjusted at 80% humidity

Signal Factor

Time was considered as the signal factor with the following levels.

M_1 = five days
M_2 = six days
M_3 = seven days

Currently, products are shipped on the seventh day.

Layout and Results

Control factors were assigned to an L_{18} array. The weight on the fifth, sixth and seventh days was measured then converted into the growth rate in a natural logarithm scale. Table 2 shows the results.

Table 2: Results of Experiment and Transformation

Day		Growth Rate			After Transformation		
		5th	6th	7th	5th	6th	7th
1	N_1	4.48	5.07	5.43	1.500	1.623	1.692
	N_2	5.08	5.46	5.80	1.625	1.697	1.758
2	N_1	4.34	4.80	5.20	1.468	1.569	1.649
	N_2	4.22	4.53	5.94	1.440	1.511	1.782
—	—	—	—	—	—	—	—
18	N_1	5.51	5.66	7.10	1.707	1.733	1.960
	N_2	5.62	6.51	7.01	1.726	1.873	1.947

Computation of S/N Ratios

Using the data in the logarithm scale, the S/N ratio is calculated using a zero-point proportional equation.

Total variation, S_T

$$S_T = 1.500^2 + 1.623^2 + 1.692^2 + 1.625^2 + 1.697^2 + 1.758^2 = 16.359796$$

Effective divider, r

$$r = 5^2 + 6^2 + 7^2 = 110$$

Linear equation, L

$$L_1 = 1.5 \times 5 + 1.623 \times 6 + 1.692 \times 7 = 29.081734$$

$$L_2 = 1.625 \times 5 + 1.697 \times 6 + 1.758 \times 7 = 30.616254$$

Variation of proportional term, S_β

$$S_\beta = 16.199317$$

Interaction between proportional term and noise, $S_{\beta \times N}$

$$S_{\beta \times N} = 0.017034$$

Error variation, S_e

$$S_e = S_T - S_\beta - S_{\beta \times N} = 16.359796 - 16.199317 - 0.017034 = 0.143445$$

Error variance, V_e

$$V_e = 0.143445/4 = 0.036$$

$$\text{S/N ratio} = 10 \log \frac{\frac{1}{2r}(S_\beta - V_e)}{V_e} = 3.596 \text{ dB}$$

$$\text{Sensitivity, S} = 10 \log \frac{1}{2r}(S_\beta - V_e) = -11.399 \text{ dB}$$

Tables 3 and 4 are the response tables for the S/N ratio and sensitivity, respectively.

Table 3: Response Tables of S/N Ratio (dB)

Factor	Level 1	Level 2	Level 3
A	3.116	3.413	-----
B	3.746	3.705	2.343
C	2.785	3.180	3.829
D	3.103	3.908	2.783
E	3.658	3.351	2.785
F	2.944	4.092	2.758
G	3.440	3.110	3.243

Table 4: Response Tables of Sensitivity (dB)

Factor	Level 1	Level 2	Level 3
A	–10.079	–10.352	-----
B	–11.394	–9.273	–9.782
C	–10.144	–10.269	–10.233
D	–10.160	–10.232	–10.254
E	–10.285	–10.162	–10.200
F	–10.186	–10.227	–10.234
G	–10.135	–10.181	–10.330

Optimization

The significant factors for the S/N ratio are B, D and F. The significant factors for sensitivity are A and B.

Factor B has the largest effect on both the S/N ratio and sensitivity. B_1 gives the highest S/N ratio and B_2 the highest sensitivity. But the difference between B_1 and B_2 on the S/N ratio is very small, only 0.04 dB. Therefore, B_2 was selected as the optimum level. The optimum combination was selected as

A1-B2-C3-D2-E1-F2-G1.

Table 4 shows the results of the confirmation under current and the optimum conditions.

Confirmation

Table 5 shows the summary of the confirmatory experiment.

In this study, the basic function of growth was evaluated by the S/N ratio of the growth rate using an exponential function.

Table 5: Summary of Confirmatory Experiment (dB)

	S/N Ratio	Sensitivity
Estimate of Optimum Condition	5.26	–9.24
Confirmation	5.72	–8.93
Current Condition	3.52	–11.49
Gain	2.20	–2.56

It was observed that the bean sprouts, which grew to about 10 times their original weight, were not good in taste and crispness. However, because of the improvement in sensitivity, i.e., the growth rate, the product could be shipped much earlier (four days instead of seven days) which resulted in a huge productivity increase. In addition, taste and crispness were improved because of the increase in growth rate, four days from seven days, resulting in fresher cells. The company was able to set a higher price because of this quality improvement.

This case study is contributed by Setsumi Yoshino, Sampo Kagaku Kogyo Co. – Japan.

CHAPTER EIGHTEEN 

Critical Parameter Characterization of a Xerographic Replenisher Dispenser

XEROX CORPORATION – USA

EXECUTIVE SUMMARY

BACKGROUND

The function of the replenisher dispenser is to deliver a metered mixture of toner (a fine powder used to imprint images on a document) and carrier (a type of steel shot used in xerographic development) to the developer housing of a copy machine. The right amount must be delivered depending on what is called percent area coverage (% AC), which is determined by the image the user is printing or copying. The system must perform accurate delivery as demanded by the customer's area coverage under various environmental conditions, over the design life of the copy machine. This study was conducted to optimize the replenisher dispenser for its robust performance at the earliest stage of the development process.

PROCESS

The replenisher dispenser system is comprised of three subsystems, the delivery auger, the bottle and extraction auger and the hopper and agitator. The first robust design (RD) experiment was performed to optimize the delivery auger subsystem with an L_9 orthogonal array. The robustness of the relation between % AC and mass flow rate was studied under noise conditions of material property variability. The second RD experiment optimized the bottle and extraction auger subsystem with an L_{18} orthogonal array. The relation between the auger RPM and flow rate was optimized for robustness against material property variability. The third RD experiment optimized the hopper and agitator function using an L_{18} array.

BENEFITS

- The system was divided into three subsystems and each subsystem was optimized for robustness against critical noise factors. All this was accomplished at the earliest phase, i.e., as soon as the concepts were selected in the product development process.

- A very wide range within the design space (675 design configurations within the design space) was tested for robustness using orthogonal arrays.
- Compared to the average design within the design space explored in this study, the improvements in signal-to-noise ratio were 6.35 dB, 9.05 dB and 3.75 dB for the three subsystems.
- All three subsystems were combined to make a complete replenisher dispenser. It was confirmed that the system performance is at the target and hence excellent. The study proved the effectiveness of subsystem optimization using the ideal function.
- The danger in a traditional design–build–test approach is that when the material formulation changes, the performance of the dispenser can degrade drastically, forcing a redesign late in the product development cycle. A significant amount of time and money is saved by designing a dispenser that is robust against varying toner properties up-front, avoiding the cost of retooling parts and slipping the schedule later.
- Robust design methodologies resulted in design of a replenisher dispenser that will be robust against future material formulation changes.

CASE STUDY

INTRODUCTION

The function of the replenisher dispenser (Figure 1) is to deliver a metered mixture of toner (a fine powder used to imprint images on a document) and carrier (a type of steel shot used in xerographic development) to the developer housing of a xerographic machine. The toner/carrier mixture, called replenisher, enters the dispenser through a customer-installed bottle. Two motor-driven augers are used to transport the replenisher from the bottle to the developer housing, where the toner develops a latent image on a photoreceptor belt, is transferred and then fused to paper and finally is delivered to the output tray or finishing devices.

Replenisher Dispenser

The replenisher dispenser can be described by three subfunctions: (1) extraction of replenisher from the bottle, (2) storage of replenisher in the hopper (a reservoir which allows the customer time to change the bottle when it is empty) and (3) delivery of replenisher to the developer through the dispense outlet. In this study, the system is broken down into these three subfunctions and RD methods are employed to optimize each. The elements used in each subfunction are described further in subsequent chapters. See Figure 1.

Figure 1: Replenisher Dispenser

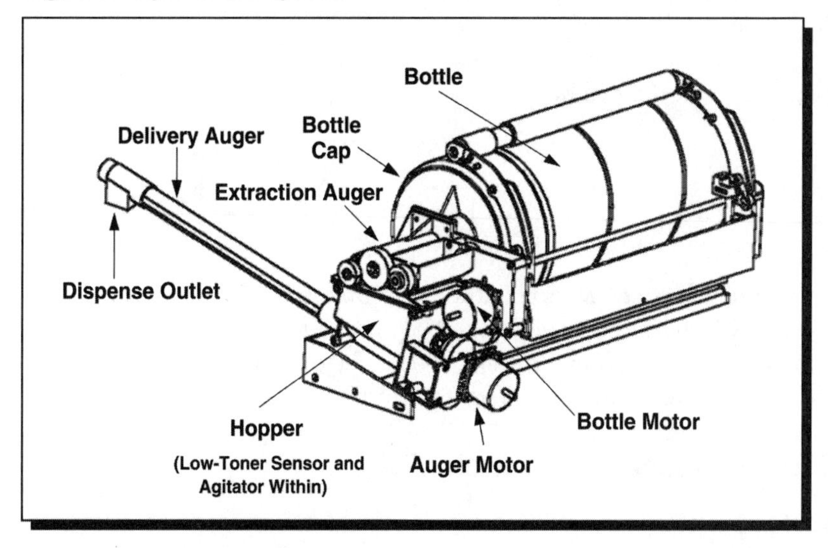

Problem Statement

The objective of using robust design methods in the design of the replenisher dispenser is to minimize the effects of varying environmental and material formulation on the performance of the system. Both environment and formulation affect the toner's ability to flow. High temperature and high humidity conditions lead to poorly flowing toner. Toner formulation, which evolves over the development cycle of a machine and beyond, affects flow based on what kind and in what concentration flow additives are used. During the parameter design phase of the product program, we cannot predict the final toner formulation which will be launched with the machine, or control the exact running environment of the machine beyond the ranges described by the system specifications. Thus the objective of the RD experiments is to minimize the effects of varying toner flow properties on each subfunction by selecting control factor settings which yield the highest S/N ratios.

Approach

Each of the three subfunctions is optimized with a separate experiment. The test is set up using a parameter diagram, and the ideal function is defined. Two popular orthogonal arrays are used: the L_9 and L_{18}. (Parameter diagrams, ideal functions and actual arrays are shown in earlier chapters where the details of the experiments are disclosed.) The L_9 and L_{18} arrays were chosen primarily because they allowed us to study three levels of control factors. In each experiment, the dynamic S/N ratio was calculated based on the following equation:

$$S/N = 10 \ \log_{10} (\beta^2 / \sigma^2)$$

where β is a regression slope of the raw data and σ^2 is the variance around that slope.

The variance is defined by

$$\sigma^2 = \frac{1}{1-n} \sum_{i=1}^{n} \Delta_i^2$$

The main effects are determined for each of the control factor settings, and the level which gives the smallest S/N is chosen for each. The improved S/N is predicted for the optimum set of control factors, and then the subfunctions are tested at those settings to verify the results.

Delivery Auger Optimization

Background

The subfunction of the delivery auger is to deliver the right amount of replenisher, measured as flow rate in units of grams per minute (g/min), to the developer. The right amount of material is determined by the customer's document, based on what is called percent area coverage (% AC), described by the image the customer is printing or copying. A simple memo, for example, would have very low % AC, on the order of 5 to 10%. A highly graphic document, such as a sales brochure, could have 85 to 90% AC depending on the size and darkness of the image. The objective is to dispense the right amount of material for the job (in g/min), regardless of material flow properties.

The geometry of the delivery auger is expected to influence how uniformly the toner is dispensed, how linear the output is as a function of auger speed, and how the dispense rate will vary with toners of differing properties. Although the theoretical volumetric output of an auger is readily calculated based on geometry, the actual output is a function of auger efficiency. The efficiency can change based on the angle of incline of the auger, the influence of the auger helix angle on the material particles and the influence of the helix angle on the material between the auger and its core diameter. The efficiency is also expected to change as a function of material flow property.

Given the number of factors which influence the performance of the auger output and the difficulty of mathematically modeling the effects of these factors, robust design methods are used to determine which factors have the most influence over variability of output.

The following auger geometry variables are dictated by the design and were treated as control factors: outside diameter, vane height (the difference between the outside diameter and the core diameter), helix angle (determined by the relationship of pitch to outside diameter, or P/D ratio) and radial clearance between the auger and the tube. The signal factor (the parameter that will drive the response) was the delivery auger speed (RPM) which will determine the % AC which the dispenser can keep up with. The material property was treated as a noise factor for this experiment, because the toner formulation changes throughout the product development cycle, and sometimes beyond. The material flow properties were treated as a noise factor and were measured as a function of the cohesiveness of the toner. The second noise factor for the experiment was the angle of incline of the auger (because at the time of the experiment, it was unclear how far uphill we would have to auger the toner). The response to be studied was the mass flow rate out of the delivery auger. The parameter diagram, illustrating the relationship of the control factors, noise factors, signal factor and response to the subfunction, is shown in Figure 2.

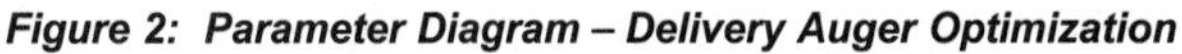
Figure 2: Parameter Diagram – Delivery Auger Optimization

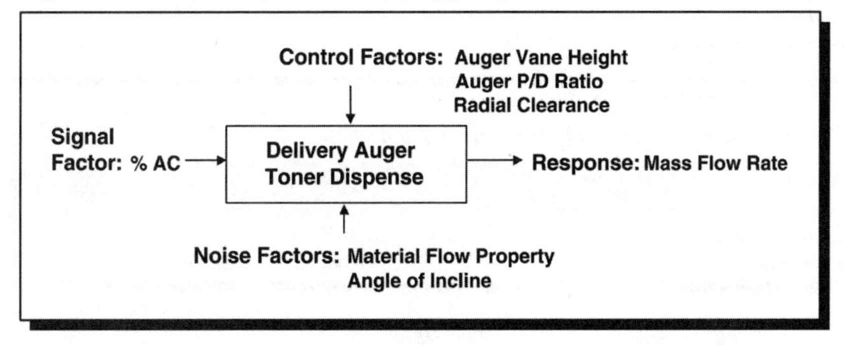

Test Objectives

The objectives of the delivery auger optimization experiment and data analysis were:

- To select control factor levels which will deliver a uniform flow rate of material from the delivery auger.
- To define the level of contribution of the vane height, radial clearance, and P/D ratio of the delivery auger on S/N and sensitivity.
- To select the optimum nominal values for the control factors which will minimize the effect of noise on the uniformity of the flow rate.
- To select control factors which would best maintain a linear relationship between the auger speed and flow rate, i.e., when the auger speed is increased by a factor of two, the flow rate should also increase by a factor of two.

Experiment

The levels selected for the control, noise and signal factors are shown in Tables 1, 2 and 3. Two levels of the material flow property noise factors were chosen: mainline toner which flows very well, and parent toner which flows very poorly. These two toners were expected to represent the best and worst case flow conditions possible. An L_9 orthogonal array was selected for the inner array, shown in Table 4, so that we could study three levels of each control factor. The response measured was the mass flow rate out of the auger. Mass measurements were recorded every 6 s. The measurements were gathered over a period of 900 s. The first 300 s were performed under the 100% area coverage condition, the next 300 s at the 50% area coverage condition, and the last 300 s at the 10% area coverage condition.

Table 1: Control Factor – Delivery Auger Optimization

Name	Level 1	Level 2	Level 3
Clearance (mm)	0.5	1.25	2
Vane Height (mm)*	3	6	9
Pitch/Diameter Ratio	1.0	0.7	0.4

*The core diameter was kept constant at 12 mm.

Table 2: Noise Factor – Delivery Auger Optimization

Name	Level 1	Level 2	Level 3
Angle of Auger	10°	20°	30°
Material	Parent	Mainline	---

Table 3: Signal Factor – Delivery Auger Optimization

Name	Level 1	Level 2	Level 3
Area Coverage	100%	50%	10%

Table 4: Orthogonal Arrary (L_9) – Delivery Auger Optimization

Cell	Left Empty	Clearance (mm)	Vane Height (mm)	P/D Ratio
1	1	0.5	3	1.0
2	1	1.25	6	0.7
3	1	2	9	0.4
4	2	0.5	6	0.4
5	2	1.25	9	1.0
6	2	2	3	0.7
7	3	0.5	9	0.7
8	3	1.25	3	0.4
9	3	2	6	1.0

Equipment

The test fixture (Figure 3) consisted of:

- Slotted mounting brackets allowing the auger tubes to be mounted at varying angles.
- An Electro-craft motor, model E286, to drive the delivery auger.
- Nine interchangeable tubes covering the range of auger diameters and radial clearances.
- Nine interchangeable augers covering the range of vane heights and pitch/diameter ratios.
- A Xerox model 4890 bottle and bottle guide used to fill the auger chamber.
- An agitator within the bottle guide to keep the material flowing between the neck of the bottle and the choke of the auger.
- A 4890 solenoid thumper used to keep the material from blocking in the bottle neck at a duty cycle of 100 ms on/2 s off.
- A Mettler PM6000 balance to weigh the toner as it is dispensed.

Figure 3: Delivery Auger Optimization Fixture

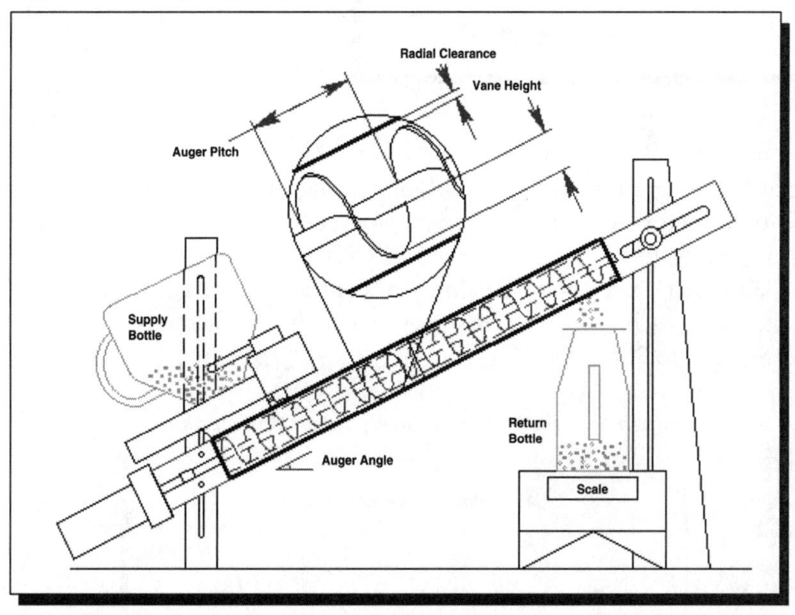

Procedure

For each test cell, the auger and tube were installed and set at a 10° angle. An empty bottle was placed on the scale, then a full bottle (containing 3 pounds of toner) was inserted in the bottle guide. The speed of the auger was set based on a given rate of 38 to 40 g/min output (100% speed). The test was run and data collected using LabVIEW®. The test was then repeated for 20° and 30°, using the same speed that was set at 10°. The mainline material was cycled through three cells of the test between material changes; the parent material was changed after every two cells.

Data Analysis

The ideal function for the delivery auger optimization (Figure 4) was:

$$\textbf{Mass} = \beta^* \textbf{(Time* \% AC)}$$

Figure 4: Delivery Auger Ideal Function

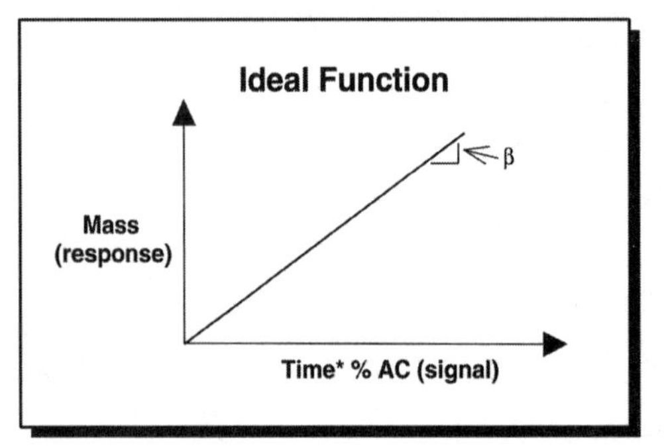

Each run provided a combination of six slopes for three angles each using the two materials. A regression analysis was performed for the experimental runs to obtain the best-fit slope and the variation around that slope. The analysis of variance (ANOVA) calculations are listed in Table 5 for all three area coverages tested.

Figure 5 illustrates the main effects of each control factor at the three different speeds. The effect of a factor level is given by the deviation of the response from the overall mean due to the factor level. The main effect plots show which factor levels are best for increasing the S/N ratio. They should give the best response with the smallest effect due to noise. The percent contribution of each control factor is listed with each curve, which shows which control factors had the largest contribution to S/N.

Figure 5: Main Effects – Delivery Auger Optimization

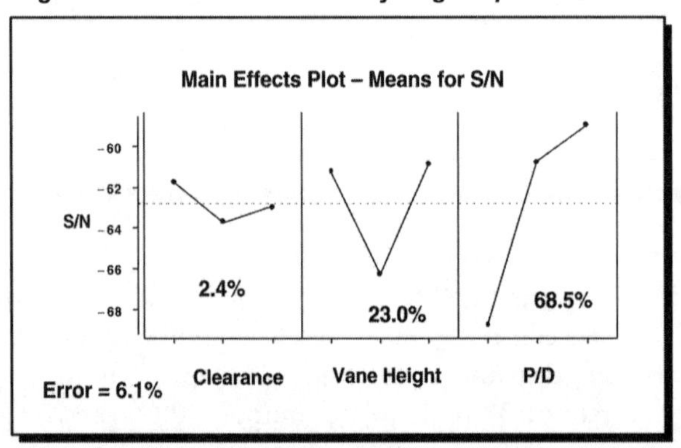

Table 5: ANOVA – Delivery Auger Optimization

Cell	β	Variance (g^2)	S/N
1	0.00628	20.0	–57.07
2	0.00587	129.5	–65.75
3	0.00531	160.0	–67.54
4	0.00476	383.7	–72.29
5	0.00563	25.5	–59.06
6	0.00604	40.2	–60.43
7	0.00611	14.8	–55.99
8	0.00553	131.3	–66.33
9	0.00613	44.9	–60.78

Source	DF	SS	MS	F
Clear	2	5.65	2.82	0.39
Vane H	2	54.39	27.19	3.76
P/D	2	162.05	81.02	11.21
Error	2	14.44	7.22	---
Total	8	236.54	---	---

DF = Degree of Freedom SS = Sum of Squares
MS = Mean Square F = F Ratio

Verification

Control factor settings were selected to maximize the signal-to-noise ratio, minimizing the sensitivity to the noise factors under study and providing a uniform flow regardless of the angle of the auger or the material properties of the toner flowing through the auger. The optimum nominal settings selected are shown in Table 6. The results of the verification are shown in Table 7, indicating a very good correlation between the predicted and actual results.

Table 6: Control Factors Selected for the Delivery Auger

Control Factors	Nominal
Vane Height	3 mm
Clearance	2 mm
P/D Ratio	1.0

Table 7: S/N Ratios for Verification – Delivery Auger

Predicted	Actual
–57.3 dB	–59.6 dB

Improvement in Signal-to-Noise Ratio

Often in robust design experiments, the level of improvement is determined by comparing the S/N ratio of the current design to the optimized design. In this case, however, there was not a design in place to compare to. The level of improvement, perhaps, can be shown by comparing the worst-case condition, which may have been selected without any experimental analysis, to the final design selected. The worst-case condition for all noise factors was –72.3 dB; the verification showed a final S/N ratio of –59.6 dB, an improvement of 12.7 dB. It is a property of the S/N ratio that for each increase of 3 dB, the variance decreases by approximately $(1/2)^{x/3}$, where x is the increase in dB. Therefore, in this experiment the variability decreased from the worst-case condition by approximately $(1/2)^{12.7/3}$, or 0.053. The level of improvement is the inverse of the reduction in variance, for an 18.8 times improvement.

Bottle/Extraction Auger Optimization

Background

The subfunction of the bottle and extraction auger subsystem is to transport material out of the bottle and into the hopper. The bottle is a customer-replaceable unit designed for easy installation and removable. There are spiral grooves molded into the sides of the cylindrically shaped bottle which, as rotated, carry material into the bottle cap. The bottle cap has within it four vanes extending radially inward from the outer diameter. The function of these vanes is to lift the material as the bottle rotates and deposit it onto the extraction auger which protrudes into the bottle. The extraction auger then carries the material out of the bottle and drops it into the hopper. The bottle and extraction auger will be driven by the same motor through a drive train. As in delivery auger optimization, the auger and bottle geometry variables are dictated by the design, and were treated as control factors. The auger geometry is defined by the outside diameter, the vane height (the difference between the outside diameter and the core diameter), the helix angle (determined by the relationship of pitch to outside diameter, or P/D ratio) and the radial clearance between the auger and the tube. The bottle geometry factors expected to influence the transportation variation were the spiral groove depth and pitch. Another factor expected to influence variability was the speed ratio between the bottle and extraction auger. The signal factor which drove the response was the extraction auger speed (RPM). The material property was treated as a noise factor for this experiment because the toner formulation changes throughout the product development cycle. The response studied was the mass flow rate out of the extraction auger. The parameter diagram, illustrating the relationship of the control factors, noise factor, signal factor, and response to the subfunction, is shown in Figure 6.

Figure 6: Parameter Diagram – Bottle/Extraction Auger Optimization

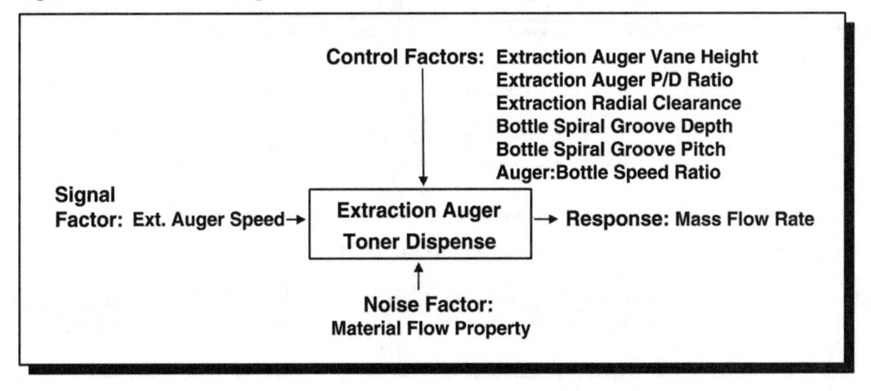

Test Objectives

The objectives of the bottle/extraction auger optimization and data analysis were:

- To select control factor levels which deliver a uniform flow rate of material from the bottle and dispense it into the hopper at a uniform rate over the life of the bottle.
- To define the level of contribution of the vane height, radial clearance, and pitch-to-diameter ratio of the extraction auger, the spiral depth and pitch of the bottle, and the speed ratio between the bottle and the auger on signal to noise and sensitivity.
- To select the optimum nominal values for the control factors which minimize the effect of noise on the uniformity of the flow rate.
- To select control factor levels which maintain a linear relationship between the auger speed and flow rate, i.e., when the auger speed is increased by a factor of two, the flow rate increases by a factor of two.

Experiment

The bottle/extraction auger optimization experiment assessed the effect of six control factors (Table 8) on the uniformity of toner delivery across two different materials: properly flowing mainline toner and poorly flowing parent toner; the outer array contained one noise factor at two levels (Table 9). Extraction auger speed was used as a signal factor at three levels (Table 10). An L_{18} orthogonal array (Table 11) was selected for the inner array.

Table 8: Control Factors – Bottle/Extraction Auger Optimization

Control Factors	Level 1	Level 2	Level 3
Radial Clearance (mm)	0.5	1.25	2
Vane Height (mm)*	3	6	9
Pitch/Diameter Ratio	1.0	0.7	0.4
Spiral Pitch (mm)	137.8	78.8	--
Spiral Depth (mm)	4	6	8
Speed Ratio	10:1	20:1	30:1

*Core diameter of the augers was kept constant at 12 mm.

Table 9: Noise Factors – Bottle/Extraction Auger Optimization

Noise Factor	Level 1	Level 2
Material	Parent	Mainline

Table 10: Signal Factors – Bottle/Extraction Auger Optimization

Signal Factor	Level 1	Level 2	Level 3
Extraction Auger RPM	30	40	50

Table 11: Orthogonal Array (L_{18}) – Bottle/Extraction Auger Optimization

Cell	Spiral Pitch (mm)	Spiral Depth (mm)	Radial Clearance (mm)	Speed Ratio	Empty	Vane Height (mm)	P/D	Empty
1	137.8	4	0.50	10:1	1	3	1.0	1
2	137.8	4	1.25	20:1	2	6	0.7	2
3	137.8	4	2.00	30:1	3	9	0.4	3
4	137.8	6	0.50	10:1	2	6	0.4	3
5	137.8	6	1.25	20:1	3	9	1.0	1
6	137.8	6	2.00	30:1	1	3	0.7	2
7	137.8	8	0.50	20:1	1	9	0.7	3
8	137.8	8	1.25	30:1	2	3	0.4	1
9	137.8	8	2.00	10:1	3	6	1.0	2
10	78.8	4	0.50	30:1	3	6	0.7	1
11	78.8	4	1.25	10:1	1	9	0.4	2
12	78.8	4	2.00	20:1	2	3	1.0	3
13	78.8	6	0.50	20:1	3	3	0.4	2
14	78.8	6	1.25	30:1	1	6	1.0	3
15	78.8	6	2.00	10:1	2	9	0.7	1
16	78.8	8	0.50	30:1	2	9	1.0	2
17	78.8	8	1.25	10:1	3	3	0.7	3
18	78.8	8	2.00	20:1	1	6	0.4	1

The response measured was mass and was recorded every 6 s. The measurements were gathered over the duration of the bottle. A LabVIEW® algorithm controlled the speed of the auger in a routine that cycled as follows: 180 s at 40 RPM, 180 s at 50 RPM, 180 s at 30 RPM. The routine repeated this progression until the bottle ran out of toner.

Equipment

The test fixture (Figure 7) consisted of. . .

- Electro-craft motors, model E286, driving a set of foam-coated rollers through an 8:1 speed reduction to drive the bottles, and the extraction augers through a 1:1 set of bevel gears.
- Six interchangeable bottles covering the range of spiral pitches and depths.
- Nine interchangeable tubes covering the range of auger diameters and clearances.
- Nine interchangeable augers varying in vane heights and pitch/diameter ratios.
- A Mettler PM6000 balance to weigh the toner as it is dispensed.

Figure 7: Bottle/Extraction Auger Optimization Fixture

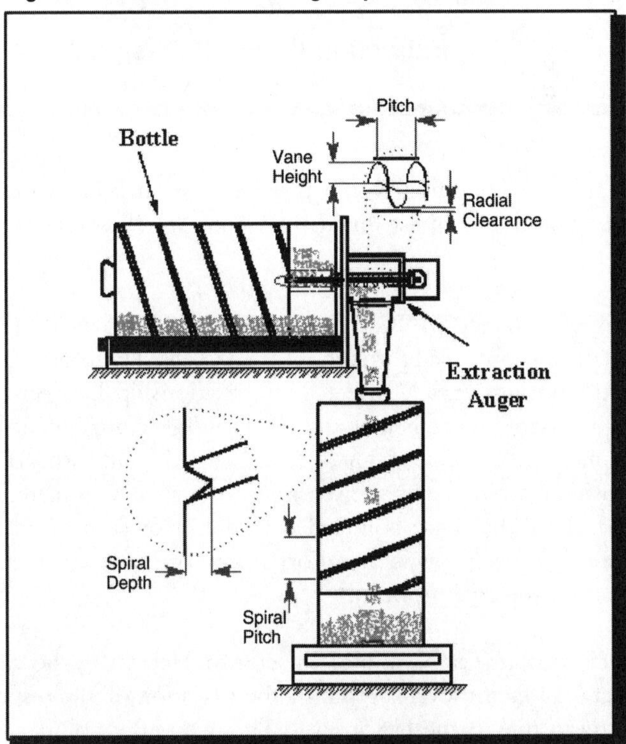

Data Analysis

The ideal function for the bottle/extraction auger optimization (Figure 8) was

$$\textbf{Flow Rate} = \beta * \textbf{RPM}$$

Each run provided experimental data of flow for all three levels of the signal factor or RPM. A regression analysis was performed for each of the experimental runs to obtain the best-fit slope and the variation around that slope. The dynamic analysis provided information on the relationship between flow

Figure 8: Bottle/Extraction Auger Ideal Function

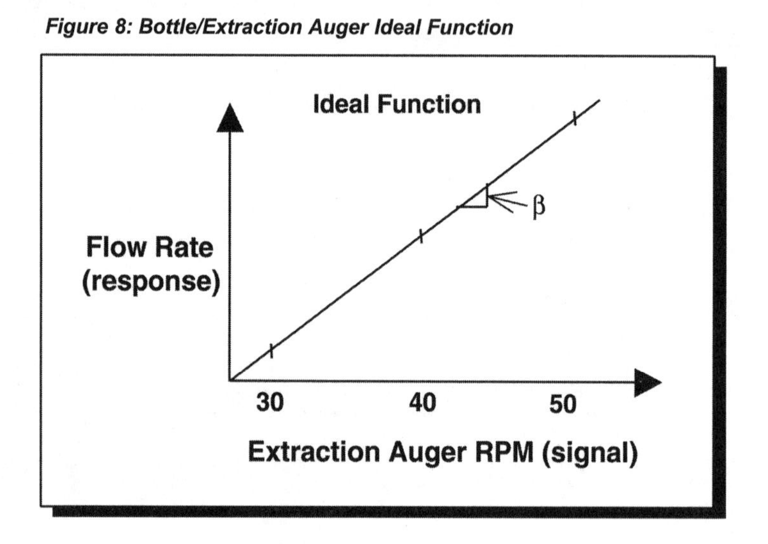

and RPM. A good condition would provide a linear relationship between RPM and flow and would also maintain a constant flow rate as the bottle empties of toner.

The analysis of variance (ANOVA) calculations are shown in Table 12. A range of 15 dB was observed throughout the 18 cells. Therefore changing the control factor settings greatly affected the relationship between RPM and flow rate; spiral depth, vane height and P/D settings contributed significantly to the change in S/N. Maximizing S/N will provide a uniform RPM to flow rate relationship by minimizing the sensitivity to a change in material properties. The effect of the noise is included in the error sum of squares. The extraction auger must provide a uniform flow of material to the hopper regardless of the material property.

The main effects of the S/N ratio of the control factors are shown in Figure 9. The effect of a factor level is given by the deviation of the response from the overall mean due to the factor level. The main effect plots show which

Table 12: ANOVA for Bottle/Extraction Auger Optimization

Cell	β	Std. Dev. (g)	S/N
1	0.0446	0.26	–15.6
2	0.1710	0.96	–15.0
3	0.1740	0.56	–10.1
4	0.0514	0.69	–22.6
5	0.4430	2.45	–14.8
6	0.0396	0.21	–14.8
7	0.3290	1.20	–11.2
8	0.0113	0.20	–25.1
9	0.1960	1.56	–18.0
10	0.1370	0.60	–12.8
11	0.1640	0.75	–13.3
12	0.0460	0.30	–16.4
13	0.0110	0.17	–24.0
14	0.1840	0.64	–10.8
15	0.3310	2.17	–16.3
16	0.4230	2.63	–15.8
17	0.0374	0.29	–17.9
18	0.0521	0.63	–21.7

Source	DF	SS	MS	F	ρ
Spiral Pitch	1	0.19	0.19	0.01	0.907
Spiral Depth	2	63.95	31.98	2.55	0.158
Clear	2	2.70	1.35	0.11	0.900
Speed Ratio	2	21.33	10.66	0.85	0.473
Vane Height	2	88.02	44.01	3.51	0.098
P/D	2	82.82	41.41	3.31	0.108
Error	6	75.18	12.53	---	---
Total	17	334.19			

DF = Degree of Freedom SS = Sum of Squares
MS = Mean Square F = F Ratio

factor levels are best for increasing the S/N ratio, which consequently decreases sensitivity to the noise factor. The percent contribution of each control factor is listed with each curve.

Table 9: S/N Main Effects – Bottle/Extraction Auger Optimization

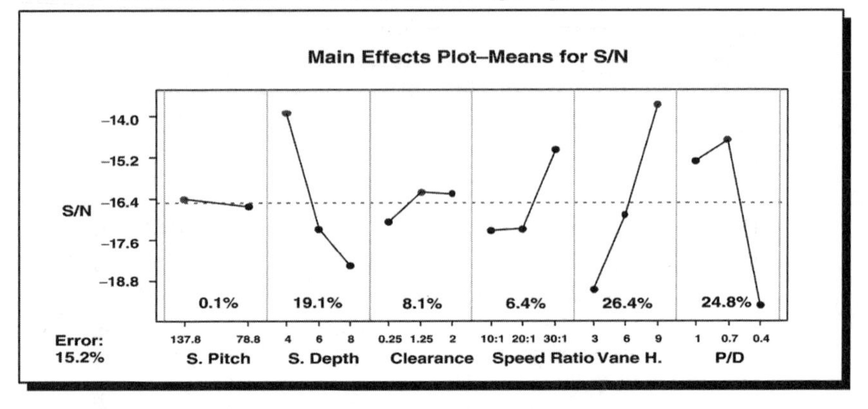

Verification

Control factor settings were selected to maximize the signal-to-noise ratio which minimizes the sensitivity to the noise factors under study. The auger speed was adjusted to 18 RPM to achieve a target flow rate of 60 g/min. The speed ratio was also adjusted to compensate for the physics of material properties at low bottle speed. A 10:1 ratio is needed to deliver the toner to the auger. The nominal settings selected are shown in Table 13. The results of the verification are shown in Table 14, indicating an improvement even greater than expected.

Table 13: Control Factors Selected for Bottle/Extraction Auger Optimization

Control Factors	Nominal
Vane Height	9 mm
P/D Ratio	0.7
Speed Ratio	10:1
Spiral Depth	4 mm
Spiral Pitch	138.7 mm
Clearance	1.25 mm

Table 14: S/N Ratio for Verification – Bottle/Extraction Auger Optimization

Predicted	Actual
–9.6 dB	–7.0 dB

Improvement in Signal-to-Noise Ratio

As in the delivery auger optimization, there was no initial design by which to compare the results. Therefore, the level of improvement is measured by comparing the worst-case S/N ratio to the final, optimized S/N ratio. Cell 8 was the worst; the S/N ratio of cell 8 was –25.1 dB. The final verification showed an S/N ratio of –7.0, an improvement of 18.1 dB. The variability was decreased by approximately $(1/2)^{18.1/3}$, or by 0.015. The level of improvement is the inverse of the reduction in variance, for a 65.5 times improvement.

HOPPER/AGITATOR OPTIMIZATION

Background

The function of the hopper in the replenisher dispenser (schematically represented in Figure 10) is to create a reservoir between the bottle and delivery auger, which provides a storage area of material to allow the customer time to change the bottle when it is emptied of replenisher. There is a sensor within the hopper that detects the level of material. When the sensor detects that the level of material has dropped below its piezoelectric blade, it sends a signal to the machine firmware which then makes a decision to turn the bottle on to refill the hopper. Every time the hopper empties enough to trigger the sensor, a count is accumulated. When the bottle is empty, that count accumulates enough hits to trigger a warning to the customer that the bottle needs to be replaced. The material remaining in the hopper allows the customer some amount of time (set by software) to replace the bottle.

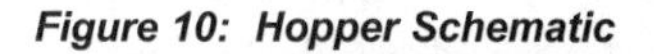

Figure 10: Hopper Schematic

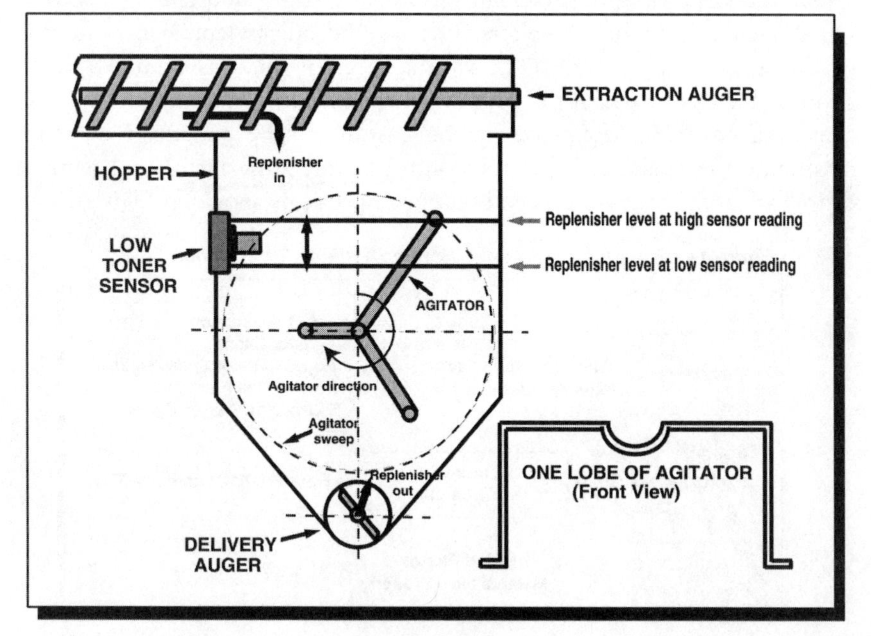

Within the hopper there is an agitator, whose function is to keep the material fluid, and also to keep the material from bridging across the sensor, which would lead to a false sensor signal (reading high when there is actually no material left in the hopper). The agitator is driven via a gear train by the same motor that drives the auger.

When material is dispensed into the hopper, it can fill with a density which may vary according to a number of different parameters. The goal of the hopper/auger optimization experiment is to determine which of these parameters affect how densely the hopper fills, and to select the parameters which will allow the amount of toner that fills the hopper to be the same regardless of the flow properties of the toner itself. One way of measuring the response of the hopper is to determine the amount of time it takes the delivery auger to deplete a given mass of toner, after the hopper has been filled by the extraction auger.

Therefore, the response to be studied in the robust design experiment is the time-out capacity of the hopper. The signal factor is the mass. The response to be measured is the countdown time from when the low toner sensor signals an empty bottle to when the hopper is empty. The control factors are items that may affect how the hopper fills. The noise factor, as in the last two experiments, is the material flow property.

For this experiment, a double-dynamic robust design experiment was used, allowing another signal factor to be introduced. The second signal factor allows the slope of the zero-point relationship to be adjusted. The second signal factor, therefore, is referred to as the adjustment signal. In this experiment, the adjustment signal was the speed of the delivery auger, which controls the rate at which the hopper is depleted, corresponding to the percent area coverage demanded by the customer. The parameter diagram, illustrating the relationship of the control factor, noise factors, adjustment factor, signal factor and measured output response, is shown in Figure 11.

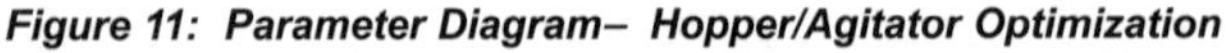

Figure 11: Parameter Diagram– Hopper/Agitator Optimization

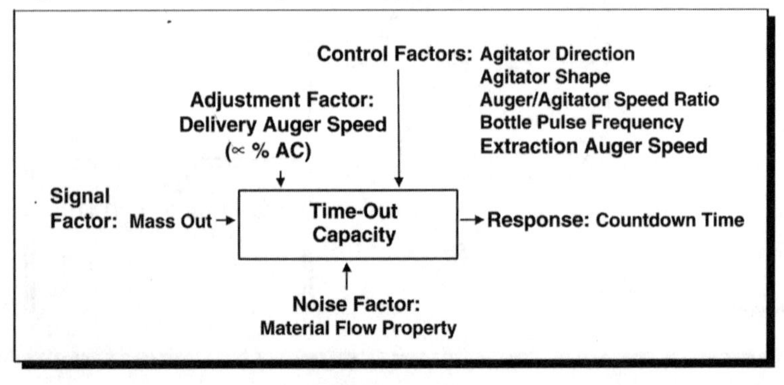

Test Objectives

The objectives of the hopper/agitator optimization experiment and data analysis were:

- To maintain a consistent time-out period from the time the low toner sensor (LTS) reads low to the time the hopper is empty over a range of material properties.
- To select the optimum control factor levels of agitator speed ratio (relative to delivery auger), agitator shape, agitator direction, hopper fill rate (i.e., extraction auger speed prior to LTS warning) and sensor polling frequency.
- To determine the level of contribution of the above control factors to the signal-to-noise ratio and sensitivity.
- To determine the relationship between time-out capacity and delivery flow rate.

Experiment

The hopper/agitator optimization experiment assessed the effect of six control factors (Table 15) on the uniformity of the rate at which the hopper empties across two materials: properly flowing mainline toner and poorly flowing parent toner; the outer array contains one noise factor at two levels (Table 16). Toner mass out was used as a signal factor (Table 17) and percent area coverage (% AC) was used as an adjustment factor (Table 18). An L_{18} orthogonal array (Table 19) was selected for the inner array. The response measured was time, in minutes, from when the sensor signal indicated low.

Table 15: Control Factor– Hopper/Agitator Optimization

Control Factors	Level 1	Level 2	Level 3
Agitator Direction	Down Toward Sensor	Up Toward Sensor	---
Agitator Shape	1 Lobe	2 Lobes	3 Lobes
Auger/Agitator Speed Ratio	1:1	2:1	3:1
Bottle Pulse Frequency	Every 3 s	Every 6 s	Every 9 s
Extraction Auger Speed	14 RPM	17 RPM	20 RPM
Speed Ratio	10:1	20:1	30:1

Table 16: Noise Factor– Hopper/Agitator Optimization

Noise	Level 1	Level 2
Material	Parent	Mainline

Table 17: Signal Factor– Hopper/Agitator Optimization

Signal
Toner Mass Out (grams)

Table 18: Adjustment Factor– Hopper/Agitator Optimization

Adjustment Factors	Level 1	Level 2	Level 3
Delivery Auger Speed	70 RPM	7 RPM	7 RPM 500 ms on and 500 ms off
Expected Output Rate	40 g/min	4 g/min	2 g/min
Corresponding AC	100%	10%	5%

Table 19: Orthogonal Array (L_{18})– Hopper/Agitator Optimization

Cell	Agitator Direction	Agitator Shape	Agitator Speed Ratio	Bottle Pulse	Ext. Auger Speed
1	Down	1 Lobe	1:1	3 s	14 RPM
2	Down	1 Lobe	2:1	6 s	17 RPM
3	Down	1 Lobe	3:1	10 s	20 RPM
4	Down	2 Lobes	1:1	3 s	17 RPM
5	Down	2 Lobes	2:1	6 s	20 RPM
6	Down	2 Lobes	3:1	10 s	14 RPM
7	Down	3 Lobes	1:1	6 s	14 RPM
8	Down	3 Lobes	2:1	10 s	17 RPM
9	Down	3 Lobes	3:1	3 s	20 RPM
10	Up	1 Lobe	1:1	10 s	20 RPM
11	Up	1 Lobe	2:1	3 s	14 RPM
12	Up	1 Lobe	3:1	6 s	17 RPM
13	Up	2 Lobes	1:1	6 s	20 RPM
14	Up	2 Lobes	2:1	10 s	14 RPM
15	Up	2 Lobes	3:1	3 s	17 RPM
16	Up	3 Lobes	1:1	10 s	17 RPM
17	Up	3 Lobes	2:1	3 s	20 RPM
18	Up	3 Lobes	3:1	6 s	14 RPM

Equipment

The test fixture consisted of

- A hopper fixture with the elements illustrated in Figure 10.
- The bottle fixture illustrated in Figure 7.
- A 12° incline delivery auger.
- A set of three interchangeable agitators (with one, two and three lobes).
- Maxon DC motors driving the delivery auger and agitator, an Electro-craft motor to drive the extraction auger and an NMB C7 stepper motor to drive the bottle.
- A Mettler PM6000 balance to weigh the toner as it is dispensed.

Procedure

LabVIEW® was programmed to execute three subroutines in the following order, based on the signal received from the low toner sensor, total mass and equivalent dispense on time (defined as the 100% delivery auger speed divided by the actual speed, multiplied by the actual elapsed time; i.e., if the 100%

speed was 70 RPM and the actual speed was 7 RPM, then each second of real time was 10 seconds of dispense on time).

Hopper fill routine: Each cell of the test started from an empty hopper and delivery auger. The program filled the hopper at the set bottle and agitator speed, while the delivery auger speed was reduced to zero, until the low toner sensor became satisfied.

Stabilization routine: This segment ensured that the system was stable at the current cell parameters. It turned on the delivery auger at the set speed, and dispensed 100 g of material at normal run conditions (bottle on only when required). The amount of time required to dispense 100 g depended on the auger speed. At 100%, it took only 2.5 min, at 10% it took 25 min, and at 5% (10% speed, pulsed at 1 Hz, 50% duty cycle) it took 50 min.

Run-out routine: Once the target mass was satisfied, the bottle was shut off (to simulate an out of toner condition) and low toner counts were accumulated until the equivalent of one-min dispense time was achieved. The program initialized a 5-min (dispense time) shutdown timer. The total delivered mass and the equivalent dispense time were recorded.

After the time-out period, the program reinitialized at the next cell conditions. A total of nine hopper fill cycles were accommodated with a single 9-pound bottle, or three cells, at the 100%, 10% and 5% speeds.

Data Analysis

The ideal function for the hopper/agitator optimization (Figure 12) was

$$\text{Time} = \beta * \text{Mass} * \%\ \text{AC}$$

Figure 12: Hopper/Agitator Ideal Function

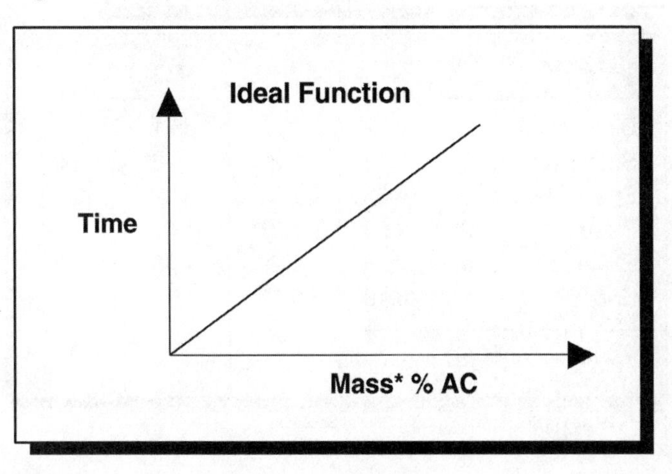

An S/N ratio was calculated for each separate run (Table 20). Each run provided experimental data for the time that lapsed to empty the hopper. A total of six data sets were combined for each run: two materials at each of the three area coverage percentages, translated into auger RPM. A regression analysis was performed on the combined data set for each experimental run to obtain the best-fit slope and the variation around each slope.

Table 20: ANOVA for Hopper/Agitator Optimization

Cell	β	Std. Dev.	S/N Ratio
1	1.68	45.50	–28.65
2	1.69	51.47	–29.67
3	1.67	50.63	–29.63
4	1.66	42.08	–28.07
5	1.65	48.23	–29.31
6	1.64	52.43	–30.09
7	1.84	43.13	–27.39
8	1.77	41.90	–27.48
9	1.76	37.10	–26.47
10	1.62	30.12	–25.38
11	1.76	34.93	–25.95
12	1.74	46.96	–28.62
13	1.55	26.14	–24.53
14	1.65	33.54	–26.16
15	1.69	34.54	–26.20
16	1.65	32.85	–25.98
17	1.72	31.84	–25.34
18	1.72	36.28	–26.48

Source	DF	Seq. SS	MS	F	ρ
Direction	1	27.2	27.2	35.1	0
Shape	2	6.46	3.23	4.18	0.05
Speed Ratio	2	4.66	2.33	3.02	0.10
Pulse	2	2.55	1.27	1.66	0.25
Ext. Auger	2	2.59	1.29	1.68	0.24
Error	8	6.18	0.77		
Total	17	49.7			

DF = Degree of Freedom SS = Sum of Squares
MS = Mean Square F = F Ratio

A range of 6 dB was observed throughout the 18 cells. Therefore, changing the control factor settings certainly affected the relationship between mass and time. The main effect plots are shown in Figure 13, including the percent contributions of each of the control factors. The direction of the agitator was by far the greatest contributor when all three speeds were analyzed together. The second largest factor was the shape of the agitator. In maximizing the S/N, the hopper is being emptied uniformly from the time the bottle goes empty (when the low toner warning is declared) and the time the hopper is empty.

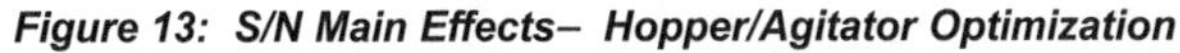

Figure 13: S/N Main Effects– Hopper/Agitator Optimization

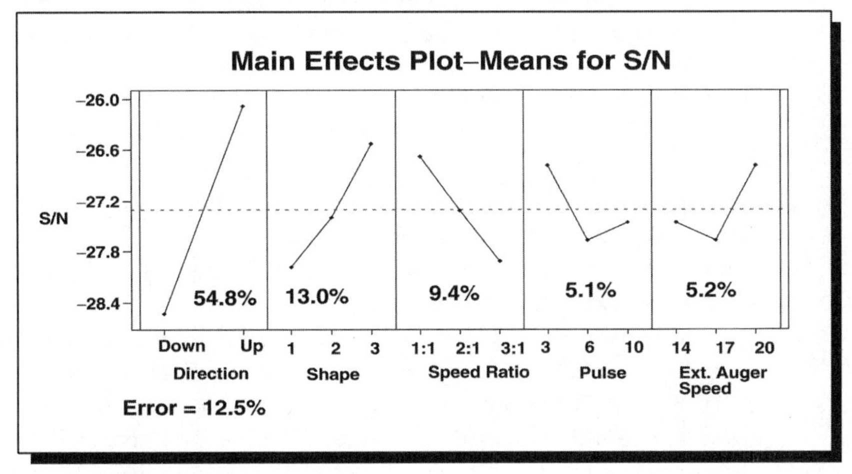

Verification

Control factors were selected to maximize the S/N ratio, which minimized the sensitivity to the noise factor under study. The nominal settings are shown in Table 21. The results of the verification are shown in Table 22, showing an actual improvement better than predicted.

Table 21: Control Factors Selected – Hopper/Agitator Optimization

Control Factors	Nominal
Direction	Up
Shape	3 Lobes
Speed Ratio	1:1
Pulse Rate	Every 3 s
Ext. Auger Speed	20 RPM

Improvement in Signal-to-Noise Ratio

As in the previous robust design experiments, there was no current design by which to quantify the level of improvement.

Table 22: S/N Ratio of Verification – Hopper/Agitator Optimization

Predicted	Actual
–23.6 dB	–22.6 dB

Therefore, the reduction of variation can be evaluated by comparing the final design selected to the worst case condition of the test. The S/N ratio of the worst-case condition was −30.1 dB. The S/N ratio of the optimized design was −22.6 dB, showing an improvement of 7.5 dB, reducing the variance by approximately $(1/2)^{7.5/3}$, or by 0.177. The level of improvement is the inverse of the reduction in variance, for a 5.6 times improvement.

Conclusion

Results Summary

- The largest contributors to the S/N ratio for the delivery auger were, in order of importance, the P/D ratio and vane height. The radial clearance had a low contribution to the S/N ratio. The optimized control factor settings for the delivery auger were a P/D of 1.0, vane height of 3 mm, and radial clearance of 2 mm.
- The largest contributor to the S/N ratio for the bottle was spiral depth. The spiral pitch had a low contribution to the S/N ratio. The optimized control factor settings for the bottle were a spiral depth of 4 mm and spiral pitch of 138.7 mm.
- The largest contributors to the S/N ratio for the extraction auger were, in order of importance, the vane height and P/D ratio. The radial clearance, speed ratio relative to the bottle, speed and pulse rate had a low contribution to the S/N ratio. The optimized control factor settings for the extraction auger were a vane height of 9 mm, P/D ratio of 0.7, radial clearance of 1.25 mm, speed ratio of 10:1 relative to the bottle, speed of 20 RPM and pulse rate of 3 s.
- The largest contributors to the S/N ratio for the agitator were, in order of importance, the direction of rotation and shape. The optimized control factor settings for the agitator were rotation direction up toward the sensor and the shape of three lobes.

System Level Verification

The design of the replenisher dispenser was completed using the control factor settings listed above. All three subsystems were combined to make a complete replenisher dispenser. The performance of the optimized system can be seen in Figure 14. The target dispense rate is shown as a solid line and the actual measured dispense rate is shown as a jagged line.

It can be seen from the graph that the dispenser output tracks very closely with the required rate. There is a slight slope to the actual output, which can be compensated for in the software algorithm which calls on the dispenser to deliver toner.

Figure 14: System-Level Verification of Replenisher Dispenser

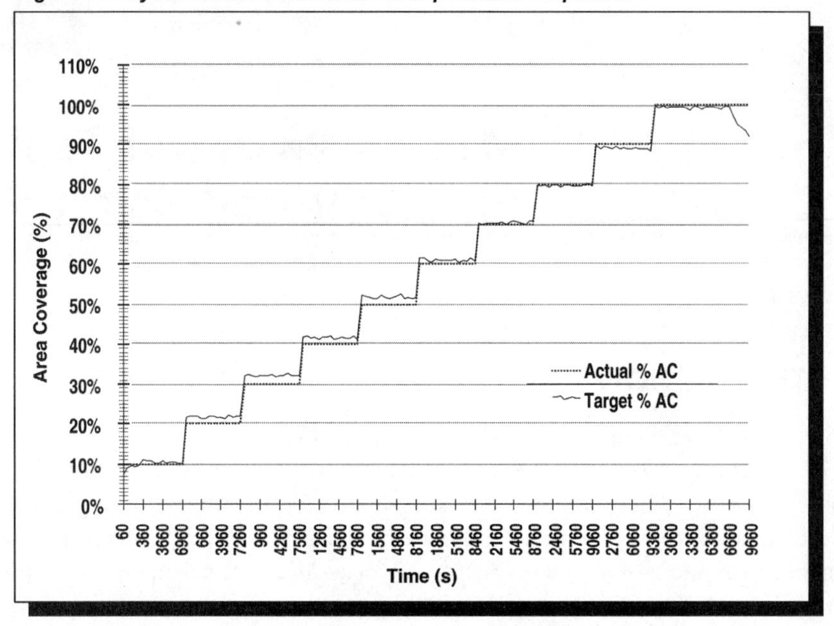

Overall, the performance of the dispenser is excellent. Historically at Xerox, replenisher dispensers have been designed to perform well with a given material. The danger in this approach is that if the material formulation changes, the performance of the dispenser can degrade drastically, forcing a redesign late in the product development cycle. A significant amount of time and money is saved by designing the dispenser to be robust against varying toner properties up-front, avoiding the cost of retooling parts and slipping the schedule later. Robust design methodologies enabled the design of a replenisher dispenser which will be robust against future material formulation changes.

This case study is contributed by Jan Enderle and Christine Keenan of Xerox Corporation, USA.

CHAPTER NINETEEN

MTS: A Powerful Tool for 2000 and Beyond

LOOKING TOWARD THE FUTURE...

INTRODUCTION

A statistical technique, Mahalanobis distance (MD), is used to distinguish one group from another considering different characteristics of a multidimensional system. The multidimensional system could be a health examination system, a university student's admission system, a printed circuit board inspection system or a sensor system. When an MD is calculated it takes care of the correlation among the characteristics of the multidimensional system. Dr. Genichi Taguchi's latest research is associated with developing robust multidimensional systems by using the Mahalanobis distance. A famous Indian statistician, P.C. Mahalanobis, the founder of the Indian Statistical Institute, introduced MD. In Japan, this method of optimization is becoming popular as the Mahalanobis–Taguchi System (MTS). There are several case studies in Japan that have demonstrated the usefulness of this method.

In quality engineering terminology, diagnosis is measurement. There are two values required in a measurement system: a base point and unit. A base point of a weighing scale is the zero point when no weight is on the scale. The unit could be a pound or a kilogram. In diagnosis applications, first we need to construct the base point (Mahalanobis space) from normal conditions. This is used as a reference point. The characteristics of multidimensional systems are compared using this reference point. If an MD corresponding to a set of characteristics is very close to MDs corresponding to the normal population, that set would be considered part of the normal population–otherwise, it is considered abnormal.

The signal-to-noise (S/N) ratio is used as a measure to compare the effectiveness of an existing system with MTS. Medical studies done in Japan under the leadership of Dr. Taguchi have demonstrated the potential of MTS to reduce cost while improving the quality of health care.

Steps in MTS

MTS can be applied to a multidimensional system in three stages. The steps in each stage are listed below.

Stage I: Construction of the Mahalanobis Space

- Define the characteristics of the multidimensional system.
- Identify the healthy (normal) population.
- Collect data or use historical data if available.
- Estimate the Mahalanobis distance by using the inverse correlation matrix. The set of MDs of the healthy group would constitute the "Mahalanobis space."
- Determine the threshold using the quality loss function.

Stage II: Diagnosis of the Existing System

- Obtain a current data set (characteristics of the multidimensional system) and compare it with the healthy group.
- Estimate the MD corresponding to this data set. If the MD is above the threshold, classify this data set as abnormal–otherwise classify it as part of the healthy group.

Stage III: Improving the Performance

- Estimate the S/N ratio of the existing system.
- Use robust design to reduce the number of characteristics needed to measure and to maintain/improve the S/N ratio. This would improve effectiveness of the system.

By using MTS one could reduce the amount of information necessary to have an effective system. To construct MS initially, it is recommended that one use as many characteristics as possible. In stage III, the robust design technique using the S/N ratio is used to determine insignificant characteristics. This would minimize the number of measurements without affecting effectiveness and resolution. This way a drastic cost reduction can be achieved in a Medicare system, in an inspection system or in an accident avoidance system.

MTS and Applications

The following areas can be considered potential candidates for the application of MTS.

- Pattern recognition for business processes
- University admission system

- System of loan/credit card approval
- Agricultural applications
- Product inspection system
- Medical diagnostic system
- Handwriting recognition system
- Manufacturing process diagnosis system
- Automobile accident avoidance system

Conclusion

- MTS will reduce the number of measurements of different characteristics into one (MD), thus helping data considerations.
- It takes care of the correlation among the characteristics of a multidimensional system.
- MD is an effective tool for differentiating a normal group with abnormalities.
- The Taguchi philosophy can be applied easily in multidimensional systems with the help of MTS.
- MTS will help eliminate the unwanted information in order to achieve the effective functioning of a system.

A

B

C

D

E

F

G

H

I

N

O

P

Q

R

U

V

W

X

Y

ABOUT THE AUTHORS

Genichi Taguchi, D.Sc., is Executive Director of American Supplier Institute, a nonprofit organization that consults on Robust Engineering. Winner of the Deming Prize of Japan and Shewhart Medal of American Society for Quality, an inductee into the Automotive Hall of Fame, and winner of many other notable awards, Dr. Taguchi developed the statistical and logical systems for rapid improvements in product quality and product development that have revolutionized automobile and other industries in America and in his home country of Japan. He is also an Honorary Member of ASQ and ASME.

Subir Chowdhury is Executive Vice President of the American Supplier Institute. He has been popularizing Robust Engineering methodology among the senior management worldwide. He is a recipient of the Society of Automotive Engineers' most prestigious Henry Ford II Award for Excellence, and was also awarded by the U.S. Congress and the Automotive Hall of Fame. He is the Chairman of the Automotive division of the American Society for Quality. He is the lead author of "Management 21C: Someday We'll All Manage This Way" (*Financial Times*, January 2000), which predicts the future of management involving the greatest management thinkers of Europe, North America, and Australia.

Shin Taguchi is President of the American Supplier Institute and the foremost consultant on Robust Engineering using Taguchi Methods in the United States. He has consulted on more than 4,000 projects for over 20 years in the Global Fortune 500 companies. He is a Fellow of the Royal Statistical Society.